GUIDE TO ESSENTIAL MATH

GUIDE TO ESSENTIAL MATH

A Review for Physics, Chemistry and Engineering Students

Second Edition

S. M. Blinder

*Wolfram Research, Inc. Champagn, IL, USA
and University of Michigan, Ann Arbor, MI, USA*

AMSTERDAM • BOSTON • HEIDELBERG • LONDON • NEW YORK • OXFORD
PARIS • SAN DIEGO • SAN FRANCISCO • SINGAPORE • SYDNEY • TOKYO

ELSEVIER

Elsevier
32 Jamestown Road, London NW1 7BY
225 Wyman Street, Waltham, MA 02451, USA

First edition 2008

Second edition 2013

Notices
Knowledge and best practice in this field are constantly changing. As new research and experience broaden our understanding, changes in research methods, professional practices, or medical treatment may become necessary.

Practitioners and researchers must always rely on their own experience and knowledge in evaluating and using any information, methods, compounds, or experiments described herein. In using such information or methods they should be mindful of their own safety and the safety of others, including parties for whom they have a professional responsibility.

To the fullest extent of the law, neither the Publisher nor the authors, contributors, or editors, assume any liability for any injury and/or damage to persons or property as a matter of products liability, negligence or otherwise, or from any use or operation of any methods, products, instructions, or ideas contained in the material herein.

British Library Cataloguing-in-Publication Data
A catalogue record for this book is available from the British Library

Library of Congress Cataloging-in-Publication Data
A catalog record for this book is available from the Library of Congress

ISBN: 978-0-323-28290-1

For information on all Elsevier publications
visit our website at store.elsevier.com

This book has been manufactured using Print On Demand technology. Each copy is produced to order and is limited to black ink. The online version of this book will show color figures where appropriate.

Working together to grow
libraries in developing countries

www.elsevier.com | www.bookaid.org | www.sabre.org

ELSEVIER   BOOK AID International   Sabre Foundation

Contents

To the Reader

Let me first tell you how the idea for this book came about. Many years ago, when I was a young assistant professor, I enthusiastically began my college teaching in a Junior-level course that was supposed to cover quantum mechanics. I was eager to share my recently-acquired insights into the intricacies of this very profound and fundamental subject which seeks to explain the structure and behavior of matter and energy. About five minutes into my first lecture, a student raised his hand and asked, "Sir, what is that funny curly thing?" The object in question was

$$\partial$$

Also, within the first week, I encountered the following very handy—but unfortunately wrong—algebraic reductions:

$$\frac{a}{x+y} = \frac{a}{x} + \frac{a}{y}, \quad \ln(x+y) = \ln x + \ln y, \quad \left(e^x\right)^2 = e^{x^2}.$$

Thus began my introduction to "Real Life" in college science courses!

All of you here—in these intermediate-level physics, chemistry, or engineering course—are obviously bright and motivated. You got through your Freshman and Sophomore years with flying colors—or at least reasonable enough success to be still here. But maybe you had a little too much fun during your early college years—which is certainly an inalienable privilege of youth! Those math courses, in particular, were often a bit on the dull side. Oh, you got through OK, maybe even with As or Bs. But somehow their content never became part of your innermost consciousness. Now you find yourself in a Junior, Senior, or Graduate level science course, with prerequisites of three or four terms of calculus. And your Prof assumes you have really mastered the stuff. On top of everything, the nice xs, ys and zs of your math courses have become ξs, ψs, ∇s and other unfriendly-looking beasts.

This is where I have come to rescue you! You don't necessarily have to go back to your prerequisite math courses. You already have, on some subconscious level, all the mathematical skills you need! So here is your handy little Rescue Manual. You can read just the parts of this book you think you need. And there are no homework assignments. Instead, we want to help you do the problems you already have in your science courses. You should, of course, work through and understand steps we have omitted in presenting important results. In many instances, it is easier to carry out a multistep derivation in your own way rather than to try and follow someone else's sequence of manipulations.

SMB
Ann Arbor
June 2006

Preface to Second Edition

I was much gratified by the reception of students and teachers to the First Edition of this Math Study Guide. In response to your many suggestions, I have made several improvements and additions. A number of Problems have been added to help solidify the student's understanding of some of the more complex concepts. I have also clarified the coverage of several topics and added a Chapter on Group Theory and a Section on Hypergeometric Functions. The figures in the Second Edition are now being rendered in full color. Finally, in the first example in Chapter 1, I have updated the format of the NCAA basketball tournament to reflect its expansion to 68 teams.

A (translated) quotation attributed to Leibniz states that "It is unworthy of excellent men to lose hours like slaves in the labor of calculation ..." With modern computer software, it is now possible to perform, with remarkable facility, not only numerical but also symbolic calculations involving algebra and calculus. I am an enthusiastic user of Mathematica™ as an indispensible aid to my mathematical and scientific work. Other symbolic mathematical programs which will provide many of the same benefits are Maple™ and Mathcad™.

A useful adjunct to this book is the Wolfram Demonstration Project. This has been available on the Web at http://demonstrations.wolfram.com since 2007 and contains a growing collection, approaching 10,000 Demonstrations, mostly on scientific and mathematical topics. This should prove very instructive to the same audience to which this book is addressed. I feel privileged to have been associated with this project. I would also like to acknowledge the assistance of my colleagues at Wolfram Research for their unfailing assistance and encouragement.

Since this little book is simply an introduction to several useful mathematical concepts, the reader will undoubtedly need to seek other sources for more exhaustive coverage of specific topics. There exist, of course, thousands of excellent textbooks and references on mathematics, but I have found it very useful to refer to two online sites as a starting point. One is Wolfram MathWorld at http://mathworld.wolfram.com. The second is the continually expanding online encyclopedia: Wikipedia, at http://en.wikipedia.org.

My sincere thanks also to my Elsevier editors, Dr. Erin Hill-Parks and Ms. Tracey Miller, for their unfailing cooperation in getting this Second Edition into production.

SMB
Ann Arbor
September 2012

1 Mathematical Thinking

Mathematics is both the queen and the handmaiden of all sciences—Eric Temple Bell.

It would be great if we could study the physical world unencumbered by the need for higher mathematics. This is probably the way it was for the ancient Greek natural philosophers. Nowadays, however, you can't get very far in science or engineering without expressing results in mathematical language. This is not entirely a bad thing, since mathematics turns out to be more than just a language for science. It can also be an inspiration for conceptual models which provide deeper insights into natural phenomena. Indeed the essence of many physical laws is most readily apparent when they are expressed as mathematical equations—think of $E = mc^2$. As you go on, you will find that mathematics can provide indispensable short-cuts for solving complicated problems. Sometimes, mathematics is actually an alternative to thinking! At the very least, mathematics is one of the greatest labor-saving inventions ever created by Mankind.

Instead of outlining some sort of "12-Step Program" for developing your mathematical skills, we will give 12 examples which will hopefully stimulate your imagination on how to think cleverly along mathematical lines. It will be worthwhile for you to study each example until you understand it completely and have internalized the line of reasoning. You might have to put off doing some of Examples 7–12 until you review some relevant calculus background in the later chapters of this book.

1.1 The NCAA March Madness Problem

Every March, according to a new format adopted in 2012, 68 college basketball teams are invited to compete in the NCAA tournament to determine a national champion. (Congratulations, 2012 Kentucky Wildcats!) Let's calculate the total number of games that have to be played. First, 8 teams participate in 4 "play in" games to reduce the field to 64, a nice power of 2. In the first round, 32 games reduce the field to 32. In the second round, 16 games produce 16 winners who move on to 4 regional tournaments. There are then 8 games followed by 4 games to determine the "final 4." The winner is then decided the following week by two semifinals followed by the championship game. If you're keeping track, the total number of games to determine the champion equals $4 + 32 + 16 + 8 + 4 + 2 + 1 = 67$. But there's a much more elegant way to

Guide to Essential Math 2e. http://dx.doi.org/10.1016/B978-0-12-407163-6.00001-1

solve the problem. Since this is a single-elimination tournament, 67 of the 68 teams have to eventually lose a game. Therefore we must play exactly 67 games!

Problem 1.1.1. The College World Series is an annual tournament (every June in Omaha, NE) to determine the college baseball champion. Eight teams are split into two, four-team, double-elimination brackets (a team must lose *two* games to be eliminated), with the winner of each bracket playing in a best-of-three championship series. Calculate the total number of games played—there are actually four possibilities.

1.2   Gauss and the Arithmetic Series

A tale in mathematical mythology—it's hard to know how close it is to the actual truth—tells of Carl Friedrich Gauss as a 10-year-old student. By one account, Gauss' math teacher wanted to take a break so he assigned a problem he thought would keep the class busy for an hour or so. The problem was to add up all the integers from 1 to 100. Supposedly, Gauss almost immediately wrote down the correct answer 5050 and sat with his hands folded for the rest of the hour. The rest of his classmates got incorrect answers! Here's how he did it. Gauss noted that the 100 integers could be arranged into 50 pairs:

1	2	3	4	5	$\cdots$	50
100	99	98	97	96	$\cdots$	51

Each pair adds up to 101, thus the sum equals $101 \times 50 = 5050$. The general result for the sum of an arithmetic progression with n terms is:

$$S_n = \frac{n}{2}(a + \ell), \tag{1.1}$$

where a and ℓ are the first and last terms, respectively. This holds true whatever the difference between successive terms.

Problem 1.2.1. Evaluate the sum $1 + 4 + 7 + 10 + 13 + 16 + 19 + 22 + 25$.

1.3   The Pythagorean Theorem

The most famous theorem in Euclidean geometry is usually credited to Pythagoras (ca. 500 BC). However, Babylonian tablets suggest that the result was known more than a thousand years earlier. The theorem states that the square of the hypotenuse c of a right triangle is equal to the sum of the squares of the lengths a and b of the other two sides:

$$a^2 + b^2 = c^2. \tag{1.2}$$

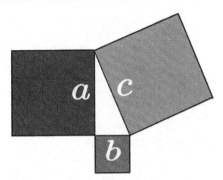

Figure 1.1 Geometrical interpretation of Pythagoras' theorem. The purple area equals the sum of the blue and red areas. (For interpretation of the references to color in this figure legend, the reader is referred to the web version of this book.)

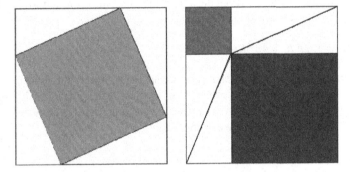

Figure 1.2 Pictorial proof of Pythagorean theorem.

The geometrical significance of the Pythagorean theorem is shown in Figure 1.1: the sum of the areas of the red and the blue squares equals the area of the purple square.

Well over 350 different proofs of the theorem have been published. Figure 1.2 shows a pictorial proof which requires neither words nor formulas.

1.4   Torus Area and Volume

A torus or anchor ring, drawn in Figure 1.3, is the approximate shape of a donut or bagel. The radii R and r refer, respectively, to the circle through the center of the torus and the circle made by a cross-sectional cut. Generally, to determine the area and volume of a surface of revolution, it is necessary to evaluate double or triple integrals. However, long before calculus was invented, Pappus of Alexandria (ca. 3rd Century AD) proposed two theorems which can give the same results much more directly.

The first theorem of Pappus states that the area A generated by the revolution of a curve about an external axis is equal to the product of the arc length of the generating

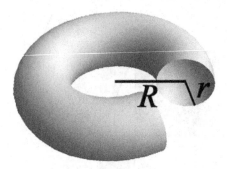

Figure 1.3 Torus. The surface of revolution with radius R of a circle of radius r.

curve and the distance traveled by the curve's centroid. For a torus the generating curve is a small circle of radius r, which has an arc length of $2\pi r$. The centroid is the center of the circle. In one revolution about the axis of the torus, the centroid travels a distance $2\pi R$. Therefore the surface area of a torus is

$$A = 2\pi r \times 2\pi R = 4\pi^2 Rr. \tag{1.3}$$

Similarly, the second theorem of Pappus states that the volume of the solid generated by the revolution of a figure about an external axis is equal to the product of the area of the figure and the distance traveled by its centroid. For a torus, the area of the cross-section equals πr^2. Therefore the volume of a torus is

$$V = \pi r^2 \times 2\pi R = 2\pi^2 Rr^2. \tag{1.4}$$

For less symmetrical figures, finding the centroid will usually require doing an integration.

   An incidental factoid. You probably know about the four-color theorem: on a plane or spherical surface, four colors suffice to draw a map in such a way that regions sharing a common boundary have different colors. On the surface of a torus it takes *seven* colors.

Problem 1.4.1.   Using the theorems of Pappus, calculate the volume and surface area of a flat cylindrical disk of width w, outer radius r_2, and inner radius r_1. Check the results using the volume and area formulas for cylinders.

1.5   Einstein's Velocity Addition Law

Suppose a baseball team is traveling on a train moving at 60 mph. The star fastball pitcher needs to tune up his arm for the next day's game. Fortunately one of the railroad cars is free and its full length is available. If his 90 mph pitches are in the same direction the train is moving, the ball will actually be moving at 150 mph relative to the ground. The law of addition of velocities in the same direction is relatively straightforward, $V = v_1 + v_2$. But according to Einstein's special theory of relativity, this is only approximately true and requires that v_1 and v_2 be small fractions of the speed of light, $c \approx 3 \times 10^8$ m/s (or 186,000 miles/s). Expressed mathematically, we can write

$$V(v_1, v_2) \approx v_1 + v_2 \quad \text{if } v_1, v_2 \ll c. \tag{1.5}$$

According to special relativity, the speed of light, when viewed from *any* frame of reference, has the same constant value c. Thus, if an atom moving at velocity v emits a light photon at velocity c, the photon will still be observed to move at velocity c, not $c + v$.

Our problem is to deduce the functional form of $V(v_1, v_2)$ consistent with these facts. It is convenient to build in the known asymptotic behavior for $v \ll c$ by defining

$$V(v_1, v_2) = f(v_1, v_2)(v_1 + v_2) \tag{1.6}$$

when $v_1 = c$, we evidently have $V = c$, so

$$f(c, v_2) = \frac{c}{c + v_2} = \frac{1}{1 + v_2/c} \tag{1.7}$$

and likewise

$$f(v_1, c) = \frac{c}{v_1 + c} = \frac{1}{1 + v_1/c}. \tag{1.8}$$

If both v_1 and v_2 equal c,

$$f(c, c) = \frac{c}{c + c} = \frac{1}{1 + 1}. \tag{1.9}$$

A few moments' reflection should convince you that a function consistent with these properties is

$$f(v_1, v_2) = \frac{1}{1 + v_1 v_2/c^2}. \tag{1.10}$$

We thus obtain Einstein's velocity addition law:

$$V = \frac{v_1 + v_2}{1 + v_1 v_2/c^2}. \tag{1.11}$$

1.6 The Birthday Problem

It is a possibly surprising fact that if you have 23 people in your class or at a party, there is a better than 50% chance that two people share the same birthday (same month and day, not necessarily in the same year). Of course, we assume from the outset that there are no sets of twins or triplets present. There are 366 possible birthdays (including February 29 for leap years). For 367 people, at least two of them would be guaranteed to have the same birthday. But you have a good probability of a match with far fewer people. Each

person in your group can have one of 366 possible birthdays. For two people, there
are 366×366 possibilities and for n people, 366^n possibilities. Now let's calculate the
number of ways that n people can have *different* birthdays. The first person can again
have 366 possible birthdays, but the second person is now limited to the remaining
365 days, the third person to 364 days, and so forth. The total number of possibilities
for no two birthdays coinciding is therefore equal to $366 \cdot 365 \cdot 364 \cdots (366 - n + 1)$.
In terms of factorials, this can be expressed $366!/(366 - n)!$. Dividing this number by
366^n then gives the probability that no two people have the same birthday. Subtracting
this from 1 gives the probability that this is *not* true, in other words, the probability that
at least two people *do* have coinciding birthdays is given by

$$P(n) = 1 - \frac{366!/(366 - n)!}{366^n}. \tag{1.12}$$

Using your calculator, you will find that for $n = 23$, $P \approx 0.506$. This means that there
is a slightly better than 50% chance that 2 out of the 23 people will have the same
birthday.

Problem 1.6.1.   For a group of just 10 people, estimate the probability that two of
their birthdays will coincide. How about a group of 100 people?

1.7   Fibonacci Numbers and the Golden Ratio

The sequence of Fibonacci numbers is given by: $0, 1, 1, 2, 3, 5, 8, 13, 21, 34, 55, 89,$
$144, 233, 377, 610, 987, \ldots$, in which each number is the sum of the two preceding
numbers. This can be expressed as

$$F_{n+1} = F_n + F_{n-1} \quad \text{with } F_0 \equiv 0, \ F_1 \equiv 1. \tag{1.13}$$

Fibonacci (whose real name was Leonardo Pisano) found this sequence as the number
of pairs of rabbits n months after a single pair begins breeding, assuming that the rabbits
produce offspring when they are two months old. As $n \to \infty$, the ratios of successive
Fibonacci numbers F_{n+1}/F_n approach a limit designated ϕ:

$$\lim_{n \to \infty} \frac{F_{n+1}}{F_n} = \phi, \tag{1.14}$$

which is equivalent to

$$\lim_{n \to \infty} \frac{F_n}{F_{n-1}} = \phi. \tag{1.15}$$

This implies the reciprocal relation

$$\lim_{n \to \infty} \frac{F_{n-1}}{F_n} = \frac{1}{\phi}. \tag{1.16}$$

Figure 1.4 The golden ratio and Leonardo da Vinci's *Mona Lisa*. The frame of the picture, as well as the rectangle outlining her face, has divine proportions.

Dividing Eq. (1.13) by F_n, we obtain

$$\frac{F_{n+1}}{F_n} = 1 + \frac{F_{n-1}}{F_n}. \tag{1.17}$$

In the limit as $n \to \infty$ this reduces to

$$\phi = 1 + \frac{1}{\phi} \tag{1.18}$$

giving the quadratic equation

$$\phi^2 - \phi - 1 = 0 \tag{1.19}$$

with roots

$$\phi = \frac{1 \pm \sqrt{5}}{2}. \tag{1.20}$$

The positive root, $\phi = (1 + \sqrt{5})/2 \approx 1.6180$, is known as the "golden ratio." According to the ancient Greeks, this was supposed to represent the most esthetically pleasing proportions for a rectangle, as shown in Figure 1.4. Leonardo da Vinci also referred to it as the "divine proportion."

Problem 1.7.1. Show that the golden ratio can be represented by the infinite sequence of nested square roots:

$$\phi = \sqrt{1 + \sqrt{1 + \sqrt{1 + \cdots}}}.$$

Calculate the approximate values of ϕ for 3, 4, and 5 square roots (and more if you use a programmable calculator or a computer program).

1.8 $\sqrt{\pi}$ in the Gaussian Integral

Laplace in 1778 proved that

$$\int_{-\infty}^{\infty} e^{-x^2} dx = \sqrt{\pi} \qquad (1.21)$$

for the definite integral of a Gaussian function. The British mathematician J.E. Littlewood judged this remarkable result as "not accessible to intuition at all." To derive Eq. (1.21), denote the integral by I and take its square:

$$I^2 = \left(\int_{-\infty}^{\infty} e^{-x^2} dx \right)^2 = \int_{-\infty}^{\infty} e^{-x^2} dx \int_{-\infty}^{\infty} e^{-y^2} dy, \qquad (1.22)$$

where the dummy variable y has been substituted for x in the last integral. The product of two integrals can be expressed as a double integral:

$$I^2 = \int_{-\infty}^{\infty} \int_{-\infty}^{\infty} e^{-(x^2+y^2)} dx \, dy. \qquad (1.23)$$

The differential $dx \, dy$ represents an element of area in Cartesian coordinates, with the domain of integration extending over the entire xy-plane. An alternative representation of the last integral can be expressed in plane polar coordinates r, θ. The two coordinate systems are related by

$$x = r \cos \theta, \quad y = r \sin \theta, \qquad (1.24)$$

so that

$$r^2 = x^2 + y^2. \qquad (1.25)$$

The element of area in polar coordinates is given by $r \, dr \, d\theta$, so that the double integral becomes

$$I^2 = \int_0^{\infty} \int_0^{2\pi} e^{-r^2} r \, dr \, d\theta. \qquad (1.26)$$

Integration over θ gives a factor 2π. The integral over r can be done after the substitution $u = r^2, du = 2r \, dr$:

$$\int_0^{\infty} e^{-r^2} r \, dr = \frac{1}{2} \int_0^{\infty} e^{-u} du = \frac{1}{2}. \qquad (1.27)$$

Therefore, $I^2 = 2\pi \times \frac{1}{2} = \pi$ and Laplace's result (1.21) follows from the square root.

Problem 1.8.1.   Evaluate the integral $\int_0^{\infty} e^{-x^2} dx$.

Problem 1.8.2.   Evaluate the integral $\int_{-\infty}^{\infty} e^{-\alpha x^2} dx$, where $\alpha > 0$.

Problem 1.8.3.   By taking the derivative with respect to α of the above integral, evaluate $\int_{-\infty}^{\infty} x^2 e^{-\alpha x^2} dx$, where $\alpha > 0$.

1.9 Function Equal to Its Derivative

The problem is to find a function $f(x)$ which is equal to its own derivative $f'(x)$. Assume that $f(x)$ can be expressed as a power series

$$f(x) = a_0 + a_1 x + a_2 x^2 + a_3 x^3 + \cdots = \sum_{n=0}^{\infty} a_n x^n. \tag{1.28}$$

Now differentiate term by term to get

$$f'(x) = a_1 + 2a_2 x + 3a_3 x^2 + \cdots = \sum_{n=1}^{\infty} n a_n x^{n-1} = \sum_{n=0}^{\infty} (n+1) a_{n+1} x^n, \tag{1.29}$$

where the second sum is obtained by the replacement $n \to n+1$. Since $f(x) = f'(x)$, we can equate the corresponding coefficients of x^n their expansions. This gives

$$a_n = (n+1) a_{n+1} \tag{1.30}$$

or

$$a_{n+1} = \frac{a_n}{n+1}, \quad n = 0, 1, 2, \ldots \tag{1.31}$$

Specifically,

$$a_1 = a_0, \quad a_2 = \frac{a_1}{2} = \frac{a_0}{1 \cdot 2}, \quad a_3 = \frac{a_2}{3} = \frac{a_0}{1 \cdot 2 \cdot 3} \cdots . \tag{1.32}$$

The constant a_0 is most conveniently set equal to 1, and so

$$a_n = \frac{1}{1 \cdot 2 \cdot 3 \cdots n} = \frac{1}{n!}. \tag{1.33}$$

Thus

$$f(x) = 1 + x + \frac{x^2}{2!} + \frac{x^3}{3!} + \cdots = \sum_{n=0}^{\infty} \frac{x^n}{n!}. \tag{1.34}$$

But this is just the well-known expansion for the exponential function, $f(x) = e^x$. The function is, as well, equal to its nth derivative: $d^n e^x / dx^n = e^x$, for all n.

A more direct way to solve the problem is to express it as a differential equation

$$\frac{df(x)}{dx} = f(x) \tag{1.35}$$

or

$$\frac{1}{f(x)} \frac{df(x)}{dx} = \frac{d}{dx} \ln f(x) = 1. \tag{1.36}$$

This can be integrated to give

$$\ln f(x) = x + \text{const.} \tag{1.37}$$

The constant can most simply be chosen equal to zero. Exponentiating the last equation—that is, taking e to the power of each side—we find again

$$e^{\ln f(x)} = f(x) = e^x. \tag{1.38}$$

1.10 Stirling's Approximation for $N!$

Recall the definition of N factorial:

$$N! = N(N-1)(N-2)\cdots 2\cdot 1. \tag{1.39}$$

Taking the natural logarithm, the product reduces to a sum:

$$\ln(N!) = \ln N + \ln(N-1) + \ln(N-2) + \cdots + \ln 2 + \ln 1 = \sum_{n=1}^{N} \ln n. \tag{1.40}$$

For large values of N, the sum over n can be well approximated by an integral over a continuous variable x:

$$\sum_{n=1}^{N} \ln n \approx \int_{0}^{N} \ln x\, dx. \tag{1.41}$$

The precise choice of lower limit is unimportant since its contribution will be negligible compared to the upper limit. The integral of the natural log is given by

$$\int \ln x\, dx = x \ln x - x, \tag{1.42}$$

which can be checked by taking the derivative. Therefore

$$\ln(N!) \approx \int_{0}^{N} \ln x\, dx = N \ln N - N. \tag{1.43}$$

This result is the logarithmic version of Stirling's approximation, which finds application in statistical mechanics.

The corresponding result for $N!$ itself, $N! = (N/e)^N$, is much less accurate, since taking exponentials magnifies the error. A more advanced version of Stirling's approximation is given by

$$N! \approx \sqrt{2\pi N}\left(\frac{N}{e}\right)^{N}. \tag{1.44}$$

Problem 1.10.1. Calculate 10! and 100! and compare the exact results with those obtained using Stirling's approximation.

1.11 Potential and Kinetic Energies

This is really a topic in elementary mechanics, but it has the potential to increase your mathematical acuity as well. Let us start with the definition of work. If a constant force F applied to a particle moves it a distance x. The work done on the particle equals $w = F \times x$. If force varies with position, the work is given by an integral over x:

$$w = \int F(x)dx. \tag{1.45}$$

For now, let us assume that the force and motion are exclusively in the x-direction. Now work has dimensions of energy and, more specifically, the energy imparted to the particle comes at the expense of the *potential energy* $V(x)$ possessed by the particle. We have therefore that

$$V(x) = -\int F(x)dx. \tag{1.46}$$

Expressed as its differential equivalent, this gives the relation between force and potential energy:

$$F(x) = -\frac{dV}{dx}. \tag{1.47}$$

The kinetic energy of a particle can be deduced from Newton's second law

$$F = ma = m\frac{dv}{dt} = m\frac{d^2x}{dt^2}, \tag{1.48}$$

where m is the mass of the particle and a, its acceleration. Acceleration is the time derivative of the velocity v or the second time derivative of the position x. In the absence of friction and other nonconservative forces, the work done on a particle, considered above, serves to increase its *kinetic energy* T—the energy a particle possesses by virtue of its motion. By the conservation of energy,

$$E = T(t) + V(t) = \text{const.} \tag{1.49}$$

The time derivatives of the energy components must therefore obey

$$\frac{dT}{dt} = -\frac{dV}{dt} = \frac{dw}{dt}. \tag{1.50}$$

From the differential relation for work

$$dw = F\,dx, \tag{1.51}$$

we obtain the work done per unit time, known as the *power*

$$\frac{dw}{dt} = F\frac{dx}{dt} = Fv. \tag{1.52}$$

Therefore, using Newton's second law

$$\frac{dT}{dt} = Fv = m\frac{dv}{dt}v = \frac{d}{dt}\left(\frac{1}{2}mv^2\right). \tag{1.53}$$

If we choose the constant of integration such that $T = 0$ when $v = 0$, we obtain the well-known equation for the kinetic energy of a particle:

$$T = \frac{1}{2}mv^2. \tag{1.54}$$

Generalizing the preceding formulas to three dimensions, work would be represented by a line integral

$$w = \int \mathbf{F} \cdot d\mathbf{r} \tag{1.55}$$

and the force would equal the negative gradient of the potential energy

$$\mathbf{F} = -\nabla V. \tag{1.56}$$

In three-dimensional vector form, Newton's second law becomes

$$\mathbf{F} = m\mathbf{a} = m\frac{d\mathbf{v}}{dt} = m\frac{d^2\mathbf{r}}{dt^2}. \tag{1.57}$$

The kinetic energy would still work out to $\frac{1}{2}mv^2$ but now

$$v^2 = \mathbf{v} \cdot \mathbf{v} = v_x^2 + v_y^2 + v_z^2. \tag{1.58}$$

Our derivation of the kinetic energy is valid in the nonrelativistic limit when $v \ll c$, the speed of light. Einstein's more general relation for the energy of a particle of mass m moving at speed v (in the absence of potential energy) is

$$E = \frac{mc^2}{\sqrt{1 - v^2/c^2}}. \tag{1.59}$$

For $v \ll c$, the square root can be expanded using the binomial theorem to give

$$E = mc^2\left(1 + \frac{1}{2}\frac{v^2}{c^2} + \cdots\right) = mc^2 + \frac{1}{2}mv^2 + \cdots. \tag{1.60}$$

The first term represents the *rest energy* of the particle—the energy equivalence of mass first understood by Einstein—while the second term reproduces our result for kinetic energy.

Problem 1.11.1. The force on a mass attached to a spring of force constant k is approximated by Hooke's law, $F = -kx$, where x is the displacement of the spring from its equilibrium length. Derive an expression for the potential energy $V(x)$.

1.12 Riemann Zeta Function and Prime Numbers

This last example is rather difficult but will really sharpen your mathematical wits if you can master it (don't worry if you can't).

The Riemann zeta function is defined by

$$\zeta(s) = 1 + \frac{1}{2^s} + \frac{1}{3^s} + \frac{1}{4^s} + \cdots = \sum_{k=1}^{\infty} \frac{1}{k^s}. \tag{1.61}$$

The function is finite for all values of s in the complex plane except for the point $s = 1$. Euler in 1737 proved a remarkable connection between the zeta function and an infinite product containing the prime numbers:

$$\zeta(s) = \left[\prod_{n=1}^{\infty} \left(1 - \frac{1}{p_n^s} \right) \right]^{-1}. \tag{1.62}$$

The product notation $\prod$ is analogous to $\sum$ for sums. More explicitly,

$$\prod_{k=1}^{n} a_k = a_1 a_2 a_3 \cdots a_n. \tag{1.63}$$

Thus

$$\prod_{n=1}^{\infty} \left(1 - \frac{1}{p_n^s} \right) = \left(1 - \frac{1}{2^s} \right) \left(1 - \frac{1}{3^s} \right) \left(1 - \frac{1}{5^s} \right) \left(1 - \frac{1}{7^s} \right) \left(1 - \frac{1}{11^s} \right) \cdots, \tag{1.64}$$

where p_n runs over all the prime numbers $2, 3, 5, 7, 11, \ldots$

To prove the result, consider first the product

$$\zeta(s) \left(1 - \frac{1}{2^s} \right) = \left(1 + \frac{1}{2^s} + \frac{1}{3^s} + \frac{1}{4^s} + \frac{1}{5^s} + \cdots \right) \left(1 - \frac{1}{2^s} \right)$$

$$= \left(1 + \frac{1}{2^s} + \frac{1}{3^s} + \frac{1}{4^s} + \frac{1}{5^s} + \frac{1}{6^s} + \cdots \right)$$

$$- \left(\frac{1}{2^s} + \frac{1}{4^s} + \frac{1}{6^s} + \frac{1}{8^s} + \cdots \right)$$

$$= 1 + \frac{1}{3^s} + \frac{1}{5^s} + \frac{1}{7^s} + \frac{1}{9^s} + \cdots. \tag{1.65}$$

The shaded terms cancel out. Multiplying by the factor $(1 - 1/2^s)$ has evidently removed all the terms divisible by 2 from the zeta function series. The next step is to multiply by $(1 - 1/3^s)$. This gives

$$
\zeta(s)\left(1 - \frac{1}{2^s}\right)\left(1 - \frac{1}{3^s}\right) = \left(1 + \frac{1}{3^s} + \frac{1}{5^s} + \frac{1}{7^s} + \frac{1}{9^s} + \cdots\right)\left(1 - \frac{1}{3^s}\right)
$$

$$
= \left(1 + \frac{1}{3^s} + \frac{1}{5^s} + \frac{1}{7^s} + \frac{1}{9^s} + \cdots\right)
$$

$$
- \left(\frac{1}{3^s} + \frac{1}{9^s} + \frac{1}{15^s} + \frac{1}{21^s} + \cdots\right)
$$

$$
= 1 + \frac{1}{5^s} + \frac{1}{7^s} + \frac{1}{11^s} + \frac{1}{13^s} + \cdots . \tag{1.66}
$$

Continuing the process, we successively remove all remaining terms containing multiples of 5, 7, 11, etc. Finally, we obtain

$$
\zeta(s) \prod_{n=1}^{\infty}\left(1 - \frac{1}{p_n^s}\right) = 1, \tag{1.67}
$$

which completes the proof of Euler's result.

1.13   How to Solve It

Although we had promised not to give you any mathematical "good advice" in abstract terms, the book by George Polya, *How to Solve It* (Princeton University Press, 1973), contains a procedural approach to problem solving which can be useful sometimes. The following summary is lifted verbatim from the book's preface.

1.13.1   *Understanding the Problem*

First. You have to understand the problem. What is the unknown? What are the data? What is the condition? Is it possible to satisfy the condition? Is the condition sufficient to determine the unknown? Or is it insufficient? Or redundant? Or contradictory? Draw a figure. Introduce suitable notation. Separate the various parts of the condition. Can you write them down?

1.13.2   *Devising a Plan*

Second. Find the connection between the data and the unknown. You may be obliged to consider auxiliary problems if an immediate connection cannot be found. You should obtain eventually a plan of the solution. Have you seen it before? Or have you seen the same problem in a slightly different form? Do you know a related problem? Do you know a theorem that could be useful? Look at the unknown! And try to think of a familiar problem having the same or a similar unknown. Here is a problem related to yours and solved before. Could you use it? Could you use its result? Could you use its method? Should you introduce some auxiliary element in order to make its use possible? Could you restate the problem? Could you restate it still differently? Go back to definitions. If you cannot solve the proposed problem try to solve first some related problem. Could you imagine a more accessible related problem? A more general problem? A more special problem? An analogous problem? Could you solve a part of the problem? Keep only a part of the condition, drop the other part; how far is the unknown then determined, how can it vary? Could you derive something useful from the data? Could you think of other data appropriate to determine the unknown? Could you change the unknown or data, or both if necessary, so that the new unknown and the new data are nearer to each other? Did you use all the data? Did you use the whole condition? Have you taken into account all essential notions involved in the problem?

1.13.3   Carrying Out the Plan

Third. Carry out your plan. Carrying out your plan of the solution, check each step. Can you see clearly that the step is correct? Can you prove that it is correct?

1.13.4   Looking Back

Fourth. Examine the solution obtained. Can you check the result? Can you check the argument? Can you derive the solution differently? Can you see it at a glance? Can you use the result, or the method, for some other problem?

1.14   A Note on Mathematical Rigor

Actually, the absence thereof. If any real mathematician should somehow come across this book, he or she will quickly recognize that it has been produced by Nonunion Labor. While I have tried very hard not to say anything grossly incorrect, I have played fast and loose with mathematical rigor. For the intended reading audience, I have made use of reasoning on a predominantly intuitive level. I am usually satisfied with making a mathematical result you need at least *plausible*, without always providing a rigorous proof. In essence, I am providing you with a lifeboat, not a luxury liner. Nevertheless, as you will see, I am not totally insensitive to mathematical beauty.

As a biologist, a physicist, and a mathematician were riding on a train moving through the countryside, all see a brown cow. The biologist innocently remarks: "I see cows are brown." The physicist corrects him: "That's too broad a generalization, you mean *some* cows are brown." Finally, the mathematician provides the definitive analysis: "There exists at least one cow, one side of which is brown."

The great physicist Hans Bethe believed in utilizing minimal mathematical complexity to solve physical problems. According to his long-time colleague E.E. Salpeter, "In his hands, approximations were not a loss of elegance but a device to bring out the basic simplicity and beauty of each field." One of our recurrent themes will be to formulate mathematical problems in a way which leads to the most transparent and direct solution, even if this has to be done at the expense of some rigor.

As the mathematician and computer scientist Richard W. Hamming opined, "Does anyone believe that the difference between the Lebesgue and Riemann integrals can have physical significance, and that whether, say, an airplane would or would not fly could depend on this difference? If such were claimed, I should not care to fly in that plane." Riemann integrals are the only type we will encounter in this book. Lebesgue integrals represent a generalization of the concept, which can deal with a larger class of functions, including those with an infinite number of discontinuities.

A somewhat irreverent spoof of how a real mathematician might rigorously express the observation "Suzy washed her face" might run as follows. Let there exist a $t_1 < 0$ such that the image of Suzy(t_1) of the point t_1 under the bijective mapping $f : t \mapsto$ Suzy(t) belongs to the set $\mathbb{G}$ of girls with dirty faces and let there exist a t_2 in the half-open interval $(t_1, 0]$ such that the image of the point t_2 under the same mapping belongs to $\mathbb{G}^C$, the complement of $\mathbb{G}$, with respect to the initial point t_1.

Just in case we have missed offending any reader, let us close with a quote from David Wells, *The Penguin Dictionary of Curious and Interesting Numbers* (Penguin Books, Middlesex, England, 1986):

> *Mathematicians have always considered themselves superior to physicists and, of course engineers. A mathematician gave this advice to his students: "If you can understand a theorem and you can prove it, publish it in a mathematics journal. If you understand it but can't prove it, submit it to a physics journal. If you can neither understand it nor prove it, send it to a journal of engineering."*

2 Numbers

Bear with me if you find some of the beginning stuff too easy. Years of bitter experience have shown that at least some of your classmates could use some brushing up at the most elementary level. You can ease yourself in at the appropriate level somewhere as we go along.

2.1 Integers

Don't laugh even if we start out in First Grade! We begin with the counting numbers: $1, 2, 3, 4, \ldots$, also known as *natural numbers*. Zero came a little later, sometime during the Middle Ages. The *integers* comprise the collection of natural numbers, their corresponding negative numbers, and zero. We can represent these compactly as the set $\mathbb{Z} = \{0, \pm 1, \pm 2, \pm 3, \ldots\}$. The mathematician Kronecker thought that: "God invented the integers, all else is the work of man." It is nevertheless possible to define the integers using various systems of axioms, two possible approaches being associated with the names *Peano's axioms* and *Zermelo-Fraenkel set theory*. Suffice to say, it will be quite adequate for our purposes to rely on intuitive notions of numbers, starting with counting on fingers and toes.

2.2 Primes

An integer divisible only by exactly two integers, itself and 1, is called a *prime number*. The number 1 is conventionally excluded since it has only one factor. Prime numbers have fascinated mathematicians and others for centuries and many of their properties remain undiscovered. As the great mathematician Euler noted "Mathematicians have tried in vain to this day to discover some order in the sequence of prime numbers, and we have reason to believe that it is a mystery into which the mind will never penetrate."

It is a good idea to know the primes less than 100, namely 2, 3, 5, 7, 11, 13, 17, 19, 23, 29, 31, 37, 41, 43, 47, 53, 59, 61, 67, 71, 73, 79, 83, 89, and 97. The *fundamental theorem of arithmetic* states that any positive integer can be represented as a product of primes in exactly one way—not counting different ordering. The primes can thus be considered the "atoms" of integers. Euclid proved that there is no largest prime—they go on forever. The proof runs as follows. Assume that there *does* exist a largest prime, say, p. Then consider the number

$$P = pp'p'' \cdots 2 + 1, \tag{2.1}$$

Guide to Essential Math 2e. http://dx.doi.org/10.1016/B978-0-12-407163-6.00002-3

where p', p'', ... are all the primes less than p. Clearly P is not divisible by any of p, p', $p'' \cdots 2$. Therefore P must also be prime with $P > p$. Thus the original assumption must have been false and there is no largest prime number.

There is no general formula or pattern for prime numbers. Their occurrence is seemingly random although they become less frequent as N increases. For large N, the number of primes $\leqslant N$ is approximated by $N/\ln N$. For example, $100/\ln(100) \approx 22$, compared with 25 actual primes < 100. The approximation gets better for larger N. An unsolved question concerns the number of "twin primes" (5 and 7, 17 and 19, 29 and 31, 41 and 43, etc.): is it finite or infinite? Another intriguing issue is the Goldbach conjecture which states that every even number greater than 2 can be written as the sum of two primes (perhaps in several ways). For example, $8 = 3 + 5$, $10 = 3 + 7$, or $5 + 5$. Nobody knows whether this conjecture is universally true or not. Neither a proof nor a counterexample has been found since 1742, when it was first proposed. It is possible that the Goldbach conjecture belongs to the category of propositions that are true but *undecidable*, in the sense of Gödel's *incompleteness theorem*.

Mersenne numbers are integers of the form $2^p - 1$, where p is prime. Written as binary numbers (see Section 2.7), they appear as strings of 1's. Mersenne numbers are prime candidates for primality, although most are not. As of 2012, 45 Mersenne primes had been identified, the largest being $2^{43112609} - 1$. This is also the largest known prime number, 12,978,189 digits long.

Why would a practical scientist or engineer be interested in mathematical playthings such as primes? Indeed several computer encryption schemes are based on the fact that finding the prime factors of a large number takes a (so far) impossibly large amount of computer time, whereas multiplying prime numbers is relatively easy. The best known algorithm is RSA (Rivest-Shamir-Adleman) public key encryption. To get a feeling for the underlying principle, factor the number 36,863. After you succeed, check your result by multiplying the two factors. Note how much faster you can do the second computation.

It has been observed that various species of cicadas reappear every 7, 13, or 17 years, periods which are prime numbers. It has been speculated that makes them statistically less vulnerable to their predators, which typically have 2-, 3-, 4-, or 6-year population cycles.

Problem 2.2.1. Factor the number 36,863, then check your result.

2.3 Divisibility

Positive integers other than 1 which are not prime are called *composite* numbers. All even integers are divisible by 2 (meaning with no remainder). A number is divisible by 3 if its digital root—the sum of its digits—is divisible by 3. Sometimes you have to do this again to the sum until you get 3, 6, or 9. For example, the digits of 31415926535897932384626434 add up to 111, whose digits, in turn, add up to 3. So the original number is divisible by 3. Divisibility by 9 works much the same way. Keep adding the digits and if you wind up with 9, then it is divisible by 9. For example,

314159265 sums to 36, thence to 9. An equivalent method is called "casting out nines." Cross out 9, likewise 6 and 3, likewise 4, 3, and 2 and so on.

An integer is divisible by 4 if its last two digits are divisible by 4. Thus 64,752 is divisible by 4 while 64,750 is not. An integer is divisible by 8 if its last *three* digits are divisible by 8. An integer is divisible by 400 if its last four digits are divisible by 400. Thus 1600 and 2000 are divisible while 1700, 1800, and 1900 are not. According to the Gregorian calendar, this determines whether the cusp year of a century is a leap year: 1900 was not but Y2K was a leap year! Divisibility by 5 is easy—any integer ending in 5 or 0. Divisibility by 10 is even easier—ask any first grader. Divisibility by 6 means an even number divisible by 3.

To determine if a number is divisible by 7, double the last digit, then subtract it from the remaining digits of the number. If you get an answer divisible by 7, then the original number is divisible by 7. You may have to apply the rule repeatedly. For example, to check whether 1316 is divisible by 7, double the last digit: $6 \times 2 = 12$, then subtract the answer from the remaining digits, giving $131 - 12 = 119$. Do it again for 119. You find $11 - 2 \times 9 = -7$. This is divisible by 7, and, therefore, so is 1316.

To test divisibility by 11, take the digits from left to right, alternately adding and subtracting. The original number is divisible by 11 if and only if the resulting number is divisible by 11. For example, for 2189, calculate $2 - 1 + 8 - 9 = 0$. This is divisible by 11, and so is 2189. There are complicated procedures for determining divisibility by 13 and 19, which we'll leave to number theorists to worry about.

Problem 2.3.1.   Factor the number 539.

2.4 Rational Numbers

A rational number is one that can be expressed as a *ratio* of two integers, say n/m with $m \neq 0$. The integers are included among the rational numbers, when n is divisible by m. Also, rational numbers have alternative forms, for example, $2/3 = 4/6 = 6/9$, etc. Let us focus on rational numbers reduced to their simplest form, with n and m *relatively prime*. Every rational number can be represented by a terminating or a periodically repeating decimal. Thus, $1/8 = 0.125$, $1/3 = 0.333333\ldots$, and $1/11 = 0.909090909\cdots$ A periodic decimal can be abbreviated using an overline on the repeating sequence, for example, $1/3 = 0.\overline{3}$ and $1/11 = 0.\overline{90}$. Given a repeating decimal, the corresponding rational number can be determined as in the following illustration. Suppose we are given $x = 0.\overline{142,857}$. Multiply this number by 1,000,000 to get

$$1,000,000x = \overline{142,857.142,857}. \tag{2.2}$$

Subtracting x, we eliminate the fractional part:

$$999,999x = 142,857, \tag{2.3}$$

so that $x = 142,857/999,999 = 1/7$.

A number that cannot be expressed as a ratio n/m is called *irrational*. (Irrational here doesn't mean "insane" but rather *not expressible as a ratio*.) The most famous irrational is $\sqrt{2}$, which did however drive the followers of Pythagoras somewhat insane. To prove the irrationality of $\sqrt{2}$, assume, to the contrary, that it *is* rational. This implies $\sqrt{2} = n/m$, where n and m are relatively prime. In particular, n and m cannot both be even numbers. Squaring, we obtain

$$n^2 = 2m^2. \tag{2.4}$$

This implies that the square of n is an even number and therefore n itself is even and can be expressed $n = p/2$, where p is another integer. This implies, in turn, that

$$m^2 = 2p^2. \tag{2.5}$$

By analogous reasoning, m must also be even. But, as we have seen, n and m cannot both be even. Therefore the assumption that $\sqrt{2}$ is rational must be false. The decimal expansion, $\sqrt{2} = 1.414213562373\ldots$, is nonterminating and nonperiodic.

2.5 Exponential Notation

Let's remind ourselves how very large and very small numbers are most conveniently written using exponential notation. The value of Avogadro's number N_A, the number of particles in a mole, is well approximated by 602,214,129,000,000,000,000,000. We obviously need a more compact notation. The key is to count how many places we move the decimal point. A decimal point moved n places to the left is represented by the exponent 10^n. Analogously, a decimal point moved n places to the right gives the exponent 10^{-n}. If the decimal place doesn't move then we understand $10^0 \equiv 1$. Thus 6 followed by 23 zeros is written 6×10^{23}. To four significant figures, Avogadro's number is given by $N_A = 6.022 \times 10^{23}$ mol^{-1}. At the other extreme is the mass of an electron $m_e = 0.00000000000000000000000000000091093\,8188$ kg. Counting 31 places to the right, we write this number much more reasonably as 9.10938×10^{-31} kg, accurate to six significant figures.

This brings us to those handy prefixes for very large and very small quantities, mainly from ancient Greek roots. One gram is a nice unit when you are considering, say, the mass of a dime, 2.268 g. For your body weight, a more convenient unit is the kilogram (kg), equal to 10^3 g (1 kg $\approx$ 2.20462 lb if you live in Brunei, Myanmar, Yemen, or the USA). Going in the other direction, 1 milligram (mg) = 10^{-3} g, 1 microgram (μg) = 10^{-6} g, 1 nanogram (ng) = 10^{-9} g, 1 picogram (pg) = 10^{-12} g, 1 femtogram (fg) = 10^{-15} g. You are unlikely to need any further prefixes (although they do go on to atto for 10^{-18}, zepto for 10^{-21}, and yocto for 10^{-24}). For some really large numbers, million 10^6 is indicated by the prefix mega (M), billion 10^9 by giga (G), trillion 10^{12} by tera (T), followed by the more exotic peta (P) for 10^{15}, exa (E) for 10^{18}, zetta (Z) for 10^{21}, and yotta (Y) for 10^{24}. We should note that our British cousins count our billions as "thousand millions," our trillions as "billions," and so on, but we all agree on the designations giga, tera, etc.

Most little kids know the sequence of doubled numbers 1, 2, 4, 8, 16, 32, 64, 128, 256, 512, 1024, ... These are, of course, successive powers of 2. Since the internal workings of computers are based on the binary number system, a memory capacity of $2^{10} = 1024$ bytes is conventionally called "1 kb" and a CPU speed of 1024 Hz is "1 kHz." Likewise multiples of $2^{20} = 1,048,576 \approx 1.05 \times 10^6$ describe 1 megabyte (MB) and 1 megahertz (MHz), while multiples of $2^{30} = 1,073,741,824 \approx 1.07 \times 10^9$ are used in the measures of 1 gigabyte (GB) and 1 gigahertz (GHz).

2.6 Powers of 10

This is a quick survey on the style of the beautiful book by Philip Morrison and Phylis Morrison, *Powers of Ten* (Scientific American Library, New York, 1994). It will give you a general feeling for the relative sizes of everything in the Universe. Almost in the middle of this logarithmic scale is the order of magnitude of human dimensions, 1 m, almost literally, "an arm's length." Two meters would be the height of a 6' 7" power forward. Moving downward, 1 cm, 1/100th of a meter is approximately the width on one of your fingernails (find out which one and you'll have a handy little ruler that you will never misplace! While you're at it, the first joint of one of your fingers should be almost exactly 1 in. long. That's handy for reading maps that say something like: "scale 1 in. = 100 miles"). 10^{-3} m is called a millimeter (mm). A dime is 1.25 mm in thickness. 10^{-6} m is called a micron (μm). It is the size of a small bacterium. 10^{-9} m is a nanometer (nm), the thickness of a layer of 5–10 atoms. This is a familiar quantity now with the rapid development of nanotechnology. Visible light has wavelength in the range of 400 nm (blue) to 700 nm (red). 10^{-10} m, called an Ångstrom unit (Å), is a useful measure of atomic dimensions, chemical bond lengths, and X-ray wavelengths. 10^{-12} m is 1 picometer (pm), sometimes a convenient measure of interatomic distances, with 1 Å = 100 pm. Nucleons (protons and neutrons) are of the order of 10^{-15} m or 1 femtometer (fm), also called 1 fermi. Electrons and other elementary particles appear to be pointlike in high-energy scattering experiments which have probed down to dimensions of the order of 10^{-18} m. Smaller dimensions are, at present, out of reach experimentally. Some "theories of everything," which seek to make quantum theory consistent with general relativity, speculate that at about 10^{-35} m (the *Planck length*), space and time themselves become granular. This is also the approximate scale of the fundamental objects in the currently popular superstring theory and *M*-theory.

Returning to the 1 m realm of everyday experience, we now go in the other direction to *larger* scales. 1 kilometer (km) = 10^3 m, about 0.62 mile, can be covered in less than a minute at highway speed. 10^6 m or 1000 km is approximately the width of Texas. 10^9 m is three times the distance to the Moon. 10^{12} m takes us out to between the orbits of Jupiter and Saturn. 10^{16} m is approximately equal to 1 light year (ly), the distance light will travel in 1 year at a speed of 3×10^8 m/s. The closest star to our Sun, Proxima Centauri, is 4.22 ly away. Our galaxy, the Milky Way, is approximately 100,000 ly (10^{21} m) in diameter. The Andromeda galaxy is approximately 2.5 million ly away ($\approx 2.5 \times 10^{19}$ m). Finally, the whole observable Universe is in the range of 10^{20} billion ly ($\approx 10^{26}$ m).

Our journey from the smallest to the largest dimensions of interest to scientists has thus covered over 60 powers of 10.

2.7 Binary Number System

Modern digital computers operate using binary logic. The computer manipulates data expressed as sequences of two possible voltage levels (commonly 0 V for binary 0 and either +3.3 V or +5 V for binary 1). In magnetic storage media, such as hard drives or floppy disks, unmagnetized dots can stand for 0's and magnetized dots for 1's.

In the familiar *decimal* or base-10 number system, a typical number such as 123.45 is explicitly represented by

$$123.45 \equiv 1 \times 10^2 + 2 \times 10^1 + 3 \times 10^0 + 4 \times 10^{-1} + 5 \times 10^{-2}. \tag{2.6}$$

The base-10 number system probably evolved from the fact that we have 10 fingers. The *binary* or base-2 number system uses just two digits, 0 and 1. It's like we use our two hands rather than our fingers to count. A binary number $b_3 b_2 b_1 b_0 \cdot b_{-1} b_{-2}$ is given by the analog of (2.6)

$$b_3 b_2 b_1 b_0 \cdot b_{-1} b_{-2} \equiv 2^3 b_3 + 2^2 b_2 + 2^1 b_1 + 2^0 b_0 + 2^{-1} b_{-1} + 2^{-2} b_{-2}. \tag{2.7}$$

Here are first few decimal numbers and their binary equivalents:

$$0 \leftrightarrow 0 \quad 1 \leftrightarrow 1 \quad 2 \leftrightarrow 10 \quad 3 \leftrightarrow 11 \quad 4 \leftrightarrow 100 \quad 5 \leftrightarrow 101$$
$$6 \leftrightarrow 110 \quad 7 \leftrightarrow 111 \quad 8 \leftrightarrow 1000 \quad 9 \leftrightarrow 1001 \quad 10 \leftrightarrow 1010. \tag{2.8}$$

Some larger binary numbers:

$$16 \leftrightarrow 1\,0000 \quad 32 \leftrightarrow 10\,0000 \quad 64 \leftrightarrow 100\,0000 \quad 128 \leftrightarrow 1000\,0000. \tag{2.9}$$

It is common practice to write binaries in groupings of four digits. The length of binary numbers might appear unwieldy compared to their decimal equivalents, but computers like them just fine.

To make binary data more compact, it is sometimes converted to *octal* notation, using the digits 0–7 to express numbers in base 8. For example, $100_{\text{oct}} = 100\,0000_{\text{bin}} = 64(\text{decimal})$. Even more compact is base-16 *hexadecimal* notation, augmenting the digits 0–9 by A, B, C, D, E, F for 10 through 15. One hex symbol corresponds very nicely to a cluster of 4 bits. For example, $F5_{\text{hex}} = 1111\,0101_{\text{bin}} = 245$. Decimal-binary-octal-hexadecimal conversions can be done on most scientific calculators.

The term *bit* is a contraction of *bi*nary dig*it*. A convenient unit in computer applications is a grouping of 8 bits, like bbbb bbbb, known as a *byte*. Each byte can represent up to 256 binary units of information. Since computers can only understand numbers, ASCII (American Standard Code for Information Interchange) is used to represent letters and other nonnumeric characters. For example, ASCII codes 65 to 90 stand for the upper-case letters A to Z, while 97 to 122 stand for lower case a to z. ASCII code 32 stands for a space. As an illustration:

Text:	R	e	s	c	u	e		G	u	i	d	e
ASCII:	82	101	115	99	117	101	32	71	117	105	100	101

Two binary numbers can be added exactly the way you did it in first grade. It's easier, in fact, since there are only four sums to remember:

$$0 + 0 = 0 \quad 0 + 1 = 1 \quad 1 + 1 = 10 \quad 1 + 1 + 1 = 11. \tag{2.10}$$

Here's the binary analog of the addition $15 + 10 = 25$:

$$\begin{array}{r} ^{1}1\ ^{1}1\ 1\ 1 \\ +\ 1\ 0\ 1\ 0 \\ \hline 1\ 1\ 0\ 0\ 1, \end{array} \tag{2.11}$$

where the "carries" are shown in red. Multiplication is also done analogously. You only need to remember that $0 \times 0 = 0$, $1 \times 0 = 0$, $1 \times 1 = 1$ and you never need to carry a 1. To multiply a binary number by 2, just write another 0 at the end.

A negative number can be represented by interpreting the leftmost bit in a byte as a $+/-$, leaving 7 bits to represent a magnitude from 0 to 127. A disadvantage of this convention is that +0 (0000 0000) is distinguished from -0 (1000 0000), which might cause difficulty when the computer is making a decision whether two values are equal. A better alternative for encoding negative quantities makes use of the 2's complement, determined as follows. First flip all the bits, replacing all 0's by 1's, and 1's by 0's. Then add 1, discarding any overflow (or carry). This is equivalent to subtracting the number from $2^8 = 256$. For example, 19 is binary 0001 0011, so -19 would be represented by 1110 1101. Constructing the 2's complement can be imagined as winding a binary odometer backward. For example, starting with 0000 0000, one click in reverse would give the reading 1111 1111, which represents -1. A second click would produce 1111 1110, which is -2.

Subtraction of a binary number is equivalent to addition of its 2's complement. For example, for signed 8-bit numbers,

$$x + (256 - y) = x - y + 256 \rightarrow x - y, \tag{2.12}$$

since 256 produces a 1 the 9th bit, which is discarded. 2's complements can be used as well in multiplication of signed binary numbers.

2.8 Infinity

The mathematical term for the size of a set is *cardinality*. The set of counting numbers $\{1, 2, 3, \ldots\}$ is never-ending and is thus *infinite* in extent, designated by the symbol ∞. (According to a children's story, the number 8 once thought it was the biggest possible number. When it found out that even larger numbers existed, it fainted. Since then, the prostrate figure "8" has been used as the symbol for infinity.) Infinity has some strange arithmetic properties including

$$\infty + 1 = \infty, \ 2 \times \infty = \infty, \ \frac{\infty}{2} = \infty, \ \sqrt{\infty} = \infty, \ \text{etc.} \tag{2.13}$$

Infinity is defined in *The Hitchhiker's Guide to the Galaxy* as "bigger than the biggest thing ever, and then some, much bigger than that, in fact, really amazingly immense ..."

The study of infinite sets was pioneered by Georg Cantor in the 19th century. One remarkable fact is that the cardinality of the counting numbers is equal to that of many of its subsets, for example, the even numbers. The reasoning is that these sets can be matched in a *one-to-one correspondence* with one another according to the following scheme:

$$
\begin{array}{ccccc}
2 & 4 & 6 & 8 & 10 \ \cdots \\
\updownarrow & \updownarrow & \updownarrow & \updownarrow & \updownarrow \ \cdots \\
1 & 2 & 3 & 4 & 5 \ \cdots
\end{array} \tag{2.14}
$$

Two sets are said to be in one-to-one correspondence if their members can be paired off in such a way that each member of the first set has exactly one counterpart in the second set and *vice versa*. Likewise, the set of odd numbers $\{1, 3, 5, 7, \ldots\}$, the set of perfect squares $\{1, 4, 9, 16, \ldots\}$, and the set of all integers $\{0, 1, -1, 2, -2, 3, -3, \ldots\}$ can be put into one-to-one correspondence with the natural numbers. Such sets are classified as *denumerable* or *denumerably infinite*. The term *countable* is also used, but generally includes finite sets as well. The cardinality of a denumerable set is represented by the symbol $\aleph_0$ (pronounced "aleph null"). A famous mathematical fable tells of the *Hilbert Hotel*, which contains a countably infinite number of rooms numbered $1, 2, 3, \ldots$ Then, whatever its occupancy, an arbitrary number of new guests—even a countably infinite number—can be accommodated by appropriately relocating the original guests, say from room number n to room number $2n$, thus freeing up all room numbers $2n - 1$.

A more general category than the integers are the *rational numbers*, ratios of integers such as n/m. The integers themselves belong to this category, for example, those cases in which $m = 1$. Although it might appear that there should be more rational numbers than integers, a remarkable fact is that the cardinality of both sets is equal. This can be shown by arranging the rational numbers in the following two-dimensional array:

$$
\begin{array}{ccccc}
1/1 & & & & \\
1/2 & 2/1 & & & \\
1/3 & 2/2 & 3/1 & & \\
1/4 & 2/3 & 3/2 & 4/1 & \\
1/5 & 2/4 & 3/3 & 4/2 & 5/1 \\
\cdots & \cdots & \cdots & &
\end{array} \tag{2.15}
$$

We can cross out entries which are "duplicates" of other entries, such as 2/2, 2/4, 3/3, 4/2, etc. This array can be "flattened" into a single list, which can then be matched in one-to-one correspondence with the natural numbers. The rationals therefore have the same cardinality, $\aleph_0$.

The cardinality of the *real* numbers involves a higher level of infinity. Real numbers are much more inclusive than rational numbers, containing as well irrational numbers such as $\sqrt{2}$ and transcendental numbers such as π and e (much more on these later). Real numbers can most intuitively be imagined as points on a line. This set of numbers or points is called the *continuum*, with a cardinality denoted by c. Following is an elegant proof by Cantor to show that c represents a *higher* order of infinity than $\aleph_0$. Let us consider just the real numbers in the interval $[0, 1]$. These can all be expressed as infinitely long decimal fractions of the form $0 \cdot abcde \ldots$ If the fraction terminates, as in the case of 0.125, we simply pad the decimal representation with an infinite number of zeros. Our (infinitely long and infinitely wide) list of real numbers can now be written

$$
\begin{aligned}
r_1 &= 0 \cdot a_1 b_1 c_1 d_1 e_1 \ldots \\
r_2 &= 0 \cdot a_2 b_2 c_2 d_2 e_2 \ldots \\
r_3 &= 0 \cdot a_3 b_3 c_3 d_3 e_3 \ldots \\
r_4 &= 0 \cdot a_4 b_4 c_4 d_4 e_4 \ldots \\
&\quad \vdots \quad \vdots \quad \vdots
\end{aligned}
\tag{2.16}
$$

Let us again presume, encouraged by our earlier successes, that we can put this set $\{r_1, r_2, r_3, \ldots\}$ into one-to-one correspondence with the set $\{1, 2, 3, \ldots\}$. But it doesn't work this time—we can find a counterexample. Focus on the digits colored in red. Let us change a_1 in the first decimal place to any digit other than a_1, change b_2 in the second decimal place to any digit other than b_2, and so on with the rest of the infinite list. The result will be a new decimal fraction that is *different* from every single real number on the list (2.16). Moreover, there is an infinite number of such exceptions. This contradicts our assumption of a one-to-one correspondence with the natural numbers. Therefore c, the cardinality of the real numbers, must represent a higher level of infinity than $\aleph_0$.

Cantor introduced higher orders of cardinality by defining *power sets*. These are sets comprising all the possible subsets of a given set. For a set with S elements, the power set is of dimension 2^S. For example, the power set of $\{1, 2, 3\}$ is made up of $2^3 = 8$ elements:

$$
\{\}, \{1\}, \{2\}, \{3\}, \{1, 2\}, \{2, 3\}, \{3, 1\}, \{1, 2, 3\}.
\tag{2.17}
$$

According to *Cantor's theorem*, a power set always has a *greater* cardinality than the original set. The power set of $\aleph_0$ is denoted by

$$
\aleph_1 = 2^{\aleph_0}.
\tag{2.18}
$$

Incidentally, $10^{\aleph_0}$ would belong to the same order of infinity. You should be able to convince yourself from the diagonal argument used above that the real numbers can be represented as a power set of the counting numbers. Therefore, we can set $c = \aleph_1$. The *continuum hypothesis*, conjectured by Cantor, states that there exists no intermediate cardinality between $\aleph_0$ and $\aleph_1$. Surprisingly, the truth of this proposition is *undecidable*. The work of Kurt Gödel and Paul Cohen showed that the hypothesis can neither be proved nor disproved using the axioms of Zermelo-Fraenkel set theory, on which the number system is based.

It suffices for most purposes for scientists and engineers to understand that the real numbers, or their geometrical equivalent, the points on a line, are *nondenumerably infinite*—meaning that they belong to a higher order of infinity than a denumerably infinite set. We thus distinguish between variables that have *discrete* and *continuous* ranges. A little free hint on English grammar, even though this is a math book: you can only have *less* of a continuous quantity (e.g. less money) but *fewer* of a discrete quantity (e.g. fewer dollars)—despite colloquial usage. Fortunately, however, you can have *more* of both!

Remarkably, the number of points on a two-dimensional, three-dimensional, or even larger continuum has the same cardinality as the number of points in one dimension. This follows from the idea that ordered pairs $\{x, y\}$, triplets $\{x, y, z\}$, and larger sets can be arranged into a single list in one-to-one correspondence with a set $\{x\}$, analogous to what was done with the rational numbers in the array (2.15).

The next higher transfinite number, assuming the continuum hypothesis is true, would be the cardinality of the power set of $\aleph_1$, namely $\aleph_2 = 2^{\aleph_1}$. This might represent the totality of possible functions or of geometrical curves.

Infinite sets must be handled with extreme caution, otherwise embarrassing paradoxes can result. Consider, for example, the following "proof" that there are no numbers greater than 2. For every number between 0 and 1 there is a corresponding number between 1 and 2—just add 1. It is also true that for every number x between 0 and 1, there is a corresponding number greater than 1, namely, its reciprocal $1/x$. Thus you can argue that every number greater than 1 must lie between 1 and 2, implying that there are no numbers greater than 2! The flaw in the preceding "proof" is the reality that every infinite set can be put into one-to-one correspondence with at least one of its proper subsets.

Problem 2.8.1. Show that the set of integers and odd-half-integers: $\{\ldots, -2, -3/2, -1, -1/2, 0, 1/2, 1, 3/2, 2, \ldots\}$ can be put into one-to-one correspondence with the natural numbers.

Problem 2.8.2. Enumerate the power set of the first four integers $\{1, 2, 3, 4\}$.

3 Algebra

3.1 Symbolic Variables

Algebra is a lot like arithmetic but deals with symbolic variables in addition to numbers. Very often these include x, y, and/or z, especially for "unknown" quantities which is often your job to solve for. Earlier letters of the alphabet such as $a, b, c, \ldots$ are often used for "constants," quantities whose values are determined by assumed conditions *before* you solve a particular problem. Most English letters find use somewhere as either variables or constants. Remember that variables are case sensitive, so X designates a different quantity than x. As the number of mathematical symbols in a technical subject proliferates, the English (really Latin) alphabet becomes inadequate to name all the needed variables. So Greek letters have to be used in addition. Here are the 24 letters of the Greek alphabet:

A, α alpha	I, ι iota	P, ρ, ϱ rho
B, β beta	K, κ kappa	Σ, σ sigma
Γ, γ gamma	Λ, λ lambda	T, τ tau
Δ, δ delta	M, μ mu	Υ, υ upsilon
E, ϵ, ε epsilon	N, ν nu	Φ, ϕ, φ phi
Z, ζ zeta	Ξ, ξ xi	X, χ chi
H, η eta	O, o omicron	Ψ, ψ psi
Θ, θ, ϑ theta	Π, π pi	Ω, ω omega

If you ever pledged a fraternity or sorority, they probably made you memorize these Greek letters (but your advantage was probably nullified by too much partying). Several upper-case Greek letters are identical to English ones, and so provide nothing new. The most famous Greek symbol is π, which stands for the universal ratio between the circumference and diameter of a circle: $\pi = 3.14159265\ldots$

The fundamental entity in algebra is an *equation*, consisting of two quantities connected by an equal sign. An equation can be thought of as a two pans of a scale, like the one held up by blindfolded Ms. Justice (Figure 3.1). When the weights of the two pans are equal, the scale is *balanced*. Weights can be added, subtracted, multiplied, or interchanged in such a way that the balance is maintained. Each such move has its analog as a legitimate algebraic operation which maintains the equality. The purpose of such operations is to get the equation into some desired form or to "solve" for one of its variables, which means to isolate it, usually on the left-hand side of the equation.

Guide to Essential Math 2e. http://dx.doi.org/10.1016/B978-0-12-407163-6.00003-5

Figure 3.1 The two sides of an equation must remain balanced just like the Scales of Justice. We will remove her blindfold to help her keep the scales balanced.

Einstein commented on algebra, "It's a merry science. When the animal we are hunting cannot be caught, we call it x temporarily and continue to hunt until it is bagged."

3.2 Legal and Illegal Algebraic Manipulations

Let's start with a simple equation

$$X + Y = ZW. \tag{3.1}$$

Following are the results of some legal things you can do

$$X = ZW - Y, \tag{3.2}$$

$$X + Y - ZW = 0, \tag{3.3}$$

$$\frac{X+Y}{Z} = W, \tag{3.4}$$

$$\frac{X+Y}{ZW} = 1, \tag{3.5}$$

$$\frac{ZW}{X+Y} = 1, \tag{3.6}$$

$$\frac{X+Y}{Z} = \frac{X}{Z} + \frac{Y}{Z}. \tag{3.7}$$

Here is a very tempting but **ILLEGAL** manipulation:

$$\frac{Z}{X+Y} = \frac{Z}{X} + \frac{Z}{Y} \quad \textbf{WRONG!} \tag{3.8}$$

A very useful reduction for ratios makes use of *crossmultiplication*:

$$\frac{X}{Y} = \frac{Z}{W} \iff XW = YZ \iff \frac{X}{Z} = \frac{Y}{W}. \tag{3.9}$$

Note that you can validly go in either direction.

For the addition and multiplication of fractions, the two key relationships are:

$$\frac{X}{Y} \times \frac{Z}{W} = \frac{XZ}{YW}, \quad \frac{X}{Y} + \frac{Z}{W} = \frac{XW + ZY}{YW}. \qquad (3.10)$$

The distributive law for multiplication states that

$$Z(X + Y) = ZX + ZY. \qquad (3.11)$$

This implies

$$(X + Y)(Z + W) = XZ + XW + YZ + YW. \qquad (3.12)$$

In particular

$$(x + y)^2 = x^2 + xy + yx + y^2 = x^2 + 2xy + y^2. \qquad (3.13)$$

Another useful relationship comes from

$$(x + y)(x - y) = x^2 - xy + yx + y^2 = x^2 - y^2. \qquad (3.14)$$

These last two formulas are worth having in your readily available memory. Complicated algebraic expressions are best handled nowadays using symbolic math programs such as *Mathematica*™.

Cancellation is a wonderful way to simplify formulas. Consider

$$\frac{abX + acY}{abZ + abW} = \frac{\boxed{a}\,bX + \boxed{a}\,cY}{\boxed{a}\,bZ + \boxed{a}\,cW} = \frac{bX + cY}{bZ + cW}, \qquad (3.15)$$

where the symbols in gray boxes are to be crossed out. But don't spoil everything by trying to cancel the bs or cs as well. The analogous cancellation can be done on the two sides of an equation:

$$\boxed{a}\,bX + \boxed{a}\,cY = \boxed{a}\,bZ + \boxed{a}\,bW \quad \Longrightarrow \quad bX + cY = bZ + cW. \qquad (3.16)$$

For your amusement, here is a "proof" that $2 = 1$. The following sequence of algebraic operations is entirely legitimate, except for one little item of trickery snuck in. Suppose we are given that

$$a = b. \qquad (3.17)$$

Then

$$a^2 = ab \qquad (3.18)$$

and

$$a^2 - b^2 = ab - b^2. \qquad (3.19)$$

Factoring both sides of the equation,

$$(a + b)(a - b) = b(a - b). \tag{3.20}$$

We can then simplify by cancellation of $(a - b)$ to get

$$a + b = b. \tag{3.21}$$

But since $a = b$ this means that $2 = 1$! Where did we go wrong?

Once you recover from shock, note that $a - b = 0$. And division by 0 is not legitimate. It is, in fact, true that $2 \times 0 = 1 \times 0$, but we can't cancel out the zeros!

3.3 Factor-Label Method

A very useful technique for converting physical quantities to alternative sets of units is the *factor-label method*. The units themselves are regarded as algebraic quantities subject to the rules of arithmetic, particularly to cancellation. To illustrate, let us calculate the speed of light in miles/s, given the metric value $c = 2.9979 \times 10^8$ m/s. First write this as an equation

$$c = \frac{2.9979 \times 10^8 \text{ m}}{1 \text{ s}}. \tag{3.22}$$

Now 1 m = 100 cm, which we can express in the form of a simple equation

$$\frac{100 \text{ cm}}{1 \text{ m}} = 1. \tag{3.23}$$

Multiplying Eq. (3.22) by 1 in the form of the last expression, and cancelling the units m from numerator and denominator, we find

$$c = \frac{2.9979 \times 10^8 \text{ m}}{1 \text{ s}} \times \frac{100 \text{ cm}}{1 \text{ m}} = 2.9979 \times 10^{10} \text{ cm/s}. \tag{3.24}$$

We continue by multiplying the result by successive factors of 1, expressed in appropriate forms, namely

$$\frac{1 \text{ in.}}{2.54 \text{ cm}} = 1, \quad \frac{1 \text{ ft}}{12 \text{ in.}} = 1, \quad \frac{1 \text{ mile}}{5280 \text{ ft}} = 1. \tag{3.25}$$

Thus we can continue our multiplication and cancellation sequence beginning with (3.24):

$$c \approx \frac{3 \times 10^8 \text{ m}}{1 \text{ s}} \times \frac{100 \text{ cm}}{1 \text{ m}} \times \frac{1 \text{ in.}}{2.54 \text{ cm}} \times \frac{1 \text{ ft}}{12 \text{ in.}} \times \frac{1 \text{ mile}}{5280 \text{ ft}}$$

$$\approx 186\,000 \text{ miles/s}, \tag{3.26}$$

a number well known to readers of science fiction. Note that singular and plural forms, e.g. "foot" and "feet," are regarded as equivalent for purposes of cancellation.

As another example, let us calculate the number of seconds in a year. Proceeding as before:

$$1 \text{ year} \times \frac{365 \text{ days}}{1 \text{ year}} \times \frac{24 \text{ h}}{1 \text{ day}} \times \frac{60 \text{ min}}{1 \text{ h}} \times \frac{60 \text{ s}}{1 \text{ min}} = 3.154 \times 10^7 \text{s}. \tag{3.27}$$

To within about 0.5%, we can approximate

$$1 \text{ year} \approx \pi \times 10^7 \text{ s}. \tag{3.28}$$

3.4 Powers and Roots

You remember of course that $x \times x = x^2$ and $x \times x \times x = x^3$, so $x^2 \times x^3 = x \times x \times x \times x \times x = x^5$. It is also easy to see that $x^3/x^2 = x$. The general formulas are:

$$x^n x^m = x^{n+m} \tag{3.29}$$

and

$$\frac{x^n}{x^m} = x^{n-m}. \tag{3.30}$$

For the case $m = n$ with $x \neq 0$, the last result implies the frequently encountered identity

$$x^0 = 1. \tag{3.31}$$

From (3.29) with $m = n$

$$x^n x^n = (x^n)^2 = x^{2n} \tag{3.32}$$

(not x^{n^2}). More generally,

$$(x^n)^m = x^{mn}. \tag{3.33}$$

Note that $x^2/x^3 = x^{-1} = 1/x$, an instance of the general result

$$x^{-n} = \frac{1}{x^n}. \tag{3.34}$$

You should be familiar with the limits as $n \to \infty$

$$\lim_{n \to \infty} x^n = \begin{cases} \infty, & \text{if } x > 1, \\ 0, & \text{if } x < 1, \end{cases} \tag{3.35}$$

and

$$\lim_{n\to\infty} x^{-n} = \begin{cases} 0 & \text{if } x > 1, \\ \infty & \text{if } x < 1. \end{cases} \tag{3.36}$$

Remember that $1/\infty = 0$ while $1/0 = \infty$. Dividing by zero sends some calculators and computers into a tizzy.

Consider a more complicated expression, say a ratio of polynomials

$$f(x) = \frac{x^2 + 2x + 3}{2x^2 + 3x + 4}.$$

In the limit as $x \to \infty$, x becomes negligible compared to x^2, as does any constant term. Therefore

$$\lim_{x\to\infty} \left[\frac{x^2 + 2x + 3}{2x^2 + 3x + 4} \right] = \frac{x^2}{2x^2} = \frac{1}{2}. \tag{3.37}$$

In the limit as $x \to 0$, on the other hand, all positive powers of x eventually become negligible compared to a constant. And so

$$\lim_{x\to 0} \left[\frac{x^2 + 2x + 3}{2x^2 + 3x + 4} \right] = \frac{3}{4}. \tag{3.38}$$

Using (3.32) with $n = m = \frac{1}{2}$ we find

$$(x^{1/2})^2 = x^1 = x. \tag{3.39}$$

Therefore $x^{1/2}$ must mean the square root of x:

$$x^{1/2} = \sqrt{x}. \tag{3.40}$$

More generally

$$x^{1/n} = \sqrt[n]{x} \tag{3.41}$$

and evidently

$$x^{m/n} = \sqrt[n]{x^m} = (\sqrt[n]{x})^m. \tag{3.42}$$

This also implies the equivalence

$$y = x^n \iff x = y^{1/n}. \tag{3.43}$$

Finally, consider the product $(xy)^2 = xy \times xy = x^2 y^2$. The general rule is

$$(xyz\ldots)^n = x^n y^n z^n \ldots \tag{3.44}$$

3.5   Logarithms

Inverse operations are pairs of mathematical manipulations in which one operation undoes the action of the other—for example, addition and subtraction, multiplication and division. The inverse of a number usually means its reciprocal, i.e. $x^{-1} = 1/x$. The product of a number and its inverse (reciprocal) equals 1. Raising to a power and extraction of a root are evidently another pair of inverse operations. An alternative inverse operation for raising to a power is taking the *logarithm*. The following relations are equivalent:

$$x = a^y \iff y = \log_a x \tag{3.45}$$

in which a is called the *base* of the logarithm.

All the formulas for manipulating logarithms can be obtained from corresponding relations involving raising to powers. If $x = a^y$ then $x^n = a^{ny}$. The last relation is equivalent to $ny = \log_a(x^n)$, therefore

$$\log(x^n) = n \log x, \tag{3.46}$$

where base a is understood. If $x = a^z$ and $y = a^w$, then $xy = a^{z+w}$ and $z + w = \log_a(xy)$. Therefore

$$\log(xy) = \log x + \log y. \tag{3.47}$$

More generally,

$$\log(x^n y^m z^p \ldots) = n \log x + m \log y + p \log z + \cdots. \tag{3.48}$$

There is no simple reduction for $\log(x + y)$—don't fall into the trap mentioned in the Preface! Since $a^1 = a$,

$$\log_a a = 1. \tag{3.49}$$

The identity $a^0 = 1$ implies that, for any base,

$$\log 1 = 0. \tag{3.50}$$

The log of a number less than 1 has a negative value. For any base $a > 1$, $a^{-\infty} = 0$, so that

$$\log 0 = -\infty. \tag{3.51}$$

To find the relationship between logarithms of different bases, suppose $x = b^y$, so $y = \log_b x$. Now, taking logs to the base a,

$$\log_a x = \log_a(b^y) = y \log_a b = \log_b x \times \log_a b. \tag{3.52}$$

In a more symmetrical form

$$\frac{\log_a x}{\log_b x} = \log_a b. \tag{3.53}$$

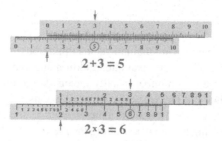

Figure 3.2 Principle of the slide rule. *Top:* Hypothetical slide rule for addition. To add 2 + 3, slide the 0 on the upper scale opposite 2 on the lower scale and look for 3 on the upper scale. The sum 5 appears below it. *Bottom:* Real slide rule based on logarithmic scale. To multiply 2 × 3, slide the 1 on the upper scale opposite 2 on the lower scale and look for 3 on the upper scale. The product 6 appears below it. Note that log 1 = 0 and that adding log 2 + log 3 gives log 6.

The slide rule, shown in Figure 3.2, is based on the principle that multiplication of two numbers is equivalent to adding their logarithms. Slide rules, once a distinguishing feature of science and engineering students, have been completely supplanted by hand-held calculators.

Logarithms to the base 10 are called *Briggsian* or *common* logarithms. Before the advent of scientific calculators, these were an invaluable aid to numerical computation. Section 1.3 on "Powers of 10" was actually a tour on the $\log_{10}$ scale. Logarithmic scales give more convenient numerical values in many scientific applications. For example, in chemistry, the hydrogen ion concentration (technically, the activity) of a solution is represented by

$$\text{pH} = -\log_{10}[\text{H}^+] \quad \text{or} \quad [\text{H}^+] = 10^{-\text{pH}}. \tag{3.54}$$

A neutral solution has a pH of 7, corresponding to $[\text{H}^+] = 10^{-7}$ moles/l. An acidic solution has pH < 7 while a basic solution has pH > 7. Another well-known logarithmic measure is the Richter scale for earthquake magnitudes. $M = 0$ is a minor tremor of some standard intensity as measured by a seismometer. The magnitude increases by 1 for every 10-fold increase in intensity. $M \geqslant 7$ is considered a "major" earthquake, capable of causing extensive destruction and loss of life. The largest magnitude in recorded history was $M = 9.5$, for the great 1960 earthquake in Chile.

Of more fundamental mathematical significance are logarithms to the base $e = 2.71828\ldots$, known as *natural logarithms*. We will explain the significance of e a little later. In most scientific usage the natural logarithm is written as "ln"

$$\log_e x \equiv \ln x. \tag{3.55}$$

But be forewarned that most literature in pure mathematics uses "log x" to mean natural logarithm. Using (3.52) with $a = e$ and $b = 10$

$$\ln x = \ln 10 \times \log_{10} x \approx 2.303 \log_{10} x. \tag{3.56}$$

Logarithms to the base 2 can be associated with the binary number system. The value of $\log_2 x$ (also written lg 2) is equal to the number of bits contained in the magnitude x. For example, $\log_2 64 = 6$.

3.6 The Quadratic Formula

The two roots of the quadratic equation

$$ax^2 + bx + c = 0 \tag{3.57}$$

are given by one of the most useful formulas in elementary algebra. We don't generally spend much time deriving formulas, but in this one instance, the derivation is very instructive. Consider the following very simple polynomial which is very easily factored:

$$x^2 + 2x + 1 = (x + 1)^2. \tag{3.58}$$

Suppose we were given instead $x^2 + 2x + 7$. We can't factor this as readily but here is an alternative trick. Knowing how the polynomial (3.58) factors we can write

$$x^2 + 2x + 7 = x^2 + 2x + 1 - 1 + 7 = (x + 1)^2 + 6. \tag{3.59}$$

This makes use of a stratagem called "completing the square." In the more general case, we can write

$$ax^2 + bx = a\left(x^2 + \frac{b}{a}x\right) = a\left[\left(x + \frac{b}{2a}\right)^2 - \frac{b^2}{4a^2}\right] = a\left(x + \frac{b}{2a}\right)^2 - \frac{b^2}{4a}. \tag{3.60}$$

This suggests how to solve the quadratic equation (3.57). First complete the square involving the first two terms:

$$ax^2 + bx + c = a\left(x + \frac{b}{2a}\right)^2 - \frac{b^2}{4a} + c = 0, \tag{3.61}$$

so that

$$\left(x + \frac{b}{2a}\right)^2 = \frac{b^2}{4a^2} - \frac{c}{a}. \tag{3.62}$$

Taking the square root:

$$x + \frac{b}{2a} = \pm\sqrt{\frac{b^2 - 4ac}{4a^2}}, \tag{3.63}$$

then leads to the famous quadratic formula

$$x = \frac{-b \pm \sqrt{b^2 - 4ac}}{2a}. \tag{3.64}$$

The quantity

$$D = b^2 - 4ac \tag{3.65}$$

is known as the *discriminant* of the quadratic equation. If $D > 0$ the equation has two distinct real roots. For example, $x^2 + x - 6$, with $D = 25$, has the two roots $x = 2$ and $x = -3$. If $D = 0$, the equation has two *equal* real roots. For example, $x^2 - 4x + 4$, with $D = 0$ has the double root $x = 2, 2$. If the discriminant $D < 0$, the quadratic formula contains the square root of a *negative* number. This leads us to *imaginary* and *complex* numbers. Before about 1800, most mathematicians would have told you that the quadratic equation with negative discriminant has *no* solutions. Associated with this point of view, the square root of a negative number has acquired the designation "imaginary." The sum of a real number with an imaginary is called a *complex number*. If we boldly accept imaginary and complex numbers, we are led to the elegant result that *every* quadratic equation has exactly two roots, whatever the sign of its discriminant. More generally, every nth-degree polynomial equation

$$a_n x^n + a_{n-1} x^{n-1} + a_{n-2} x^{n-2} + \cdots + a_0 = 0 \tag{3.66}$$

has exactly n roots.

The simplest quadratic equation with imaginary roots is

$$x^2 + 1 = 0. \tag{3.67}$$

Applying the quadratic formula (3.64), we obtain the two roots

$$x = \pm\sqrt{-1}. \tag{3.68}$$

As another example,

$$x^2 + 2x + 6 = 0 \tag{3.69}$$

has the roots

$$x = -1 \pm \sqrt{-5}. \tag{3.70}$$

Observe that whenever $D < 0$, the roots occur as *conjugate pairs*, one root containing $\sqrt{-|D|}$ and the other, $-\sqrt{-|D|}$.

The three quadratic equations considered above can be solved graphically, as shown in Figure 3.3. The two points where the parabola representing the equation crosses the x-axis correspond to the real roots. For a double root, the curve is tangent to the x-axis. If there are no real roots, as in the case of $x^2 + 1 = 0$, the curve does not intersect the x-axis.

Problem 3.6.1. Solve the quadratic equation $9x^2 - 24x + 20 = 0$.

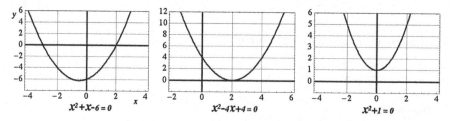

Figure 3.3 Graphical solution of quadratic equations.

3.7 Imagining i

If we an nth-degree polynomial does indeed have a total of n roots, then we must accept roots containing square roots of negative numbers—imaginary and complex numbers. The designation "imaginary" is an unfortunate accident of history since we will show that $\sqrt{-1}$ is, in fact, no more fictitious than 1 or 0—it's just a *different* kind of number, with as much fundamental significance as those we respectfully call *real numbers*.

The square root of a negative number is a multiple of $\sqrt{-1}$. For example, $\sqrt{-5} = \sqrt{5} \times \sqrt{-1}$. The *imaginary unit* is defined by

$$i \equiv \sqrt{-1}. \tag{3.71}$$

Consequently

$$i^2 = -1. \tag{3.72}$$

Clearly, there is no place on the axis of real numbers running from $-\infty$ to $+\infty$ to accommodate imaginary numbers. We are therefore forced to move into a higher dimension, representing all possible polynomial roots on a two-dimensional plane. This is known as a *complex plane* or *Argand diagram*, shown in Figure 3.4. The abscissa ("x-axis") is called the *real axis* while the ordinate ("y-axis") is called the *imaginary axis*. A quantity having both a real and an imaginary part is called a *complex number*. Every complex number is thus represented by a point on the Argand diagram. Recognize the fact that your name can be considered as a single entity, not requiring you to always spell out its

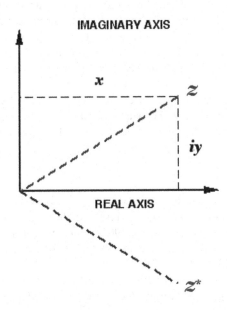

Figure 3.4 Complex plane, spanned by real and imaginary axes. The point representing $z = x + iy$ is shown along with the complex conjugate $z^* = x - iy$. Also shown is the modulus $|z|$.

individual letters. Analogously, a complex number can be considered as a single entity, commonly denoted by z, where

$$z = x + iy. \tag{3.73}$$

The real part of a complex number is denoted by $x = \Re z$ (or Re(z)) and the imaginary part by $y = \Im z$ (or Im(z)). The *complex conjugate* z^* (written $\bar{z}$ in some books) is the number obtained by changing i to $-i$:

$$z^* = x - iy. \tag{3.74}$$

As we have seen, if z is a root of a polynomial equation, then z^* is also a root. Recall that for real numbers, *absolute value* refers to the magnitude of a number, independent of its sign. Thus $|3.14| = |-3.14| = 3.14$. We can also write $-3.14 = -|3.14|$. The absolute value of a complex number z, also called its *magnitude* or *modulus*, is likewise written $|z|$. It is defined by

$$|z|^2 = zz^* = (x + iy)(x - iy) = x^2 - i^2y^2 = x^2 + y^2. \tag{3.75}$$

Thus

$$|z| = \sqrt{x^2 + y^2}, \tag{3.76}$$

which by the Pythagorean theorem is just the distance on the Argand diagram from the origin to the point representing the complex number.

The significance of imaginary and complex numbers can also be understood from a geometric perspective. Consider a number on the positive real axis, say $x > 0$. This can be transformed by a 180° rotation into the corresponding quantity $-x$ on the *negative* half of the real axis. The rotation is accomplished by multiplying x by -1. More generally, any complex number z can be rotated by 180° on the complex plane by multiplying it by -1, designating the transformation $z \to -z$. Consider now a rotation by just 90°, say counterclockwise. Let us denote this counterclockwise 90° rotation by $z \to iz$ (repress for the moment what i stands for). A second counterclockwise 90° rotation then produces the same result as a single 180° rotation. We can write this algebraically as

$$i^2z = -z, \tag{3.77}$$

which agrees with the previous definition of i in Eqs. (3.71) and (3.72). A 180° rotation following a counterclockwise 90° rotation results in the net transformation $z \to -iz$. Since this is equivalent to a single *clockwise* 90° rotation, we can interpret multiplication by $-i$ as this operation. Note that a second clockwise 90° rotation again produces the same result as a single 180° rotation. Thus $(-i)^2 = -1$ as well. Evidently $\sqrt{-1} = \pm i$. It is conventional, however, to define i as the *positive* square root of -1. Counterclockwise 90° rotation of the complex quantity $z = x + iy$ is expressed algebraically by

$$i(x + iy) = ix + i^2y = -y + ix \tag{3.78}$$

and can be represented graphically as shown in Figure 3.5.

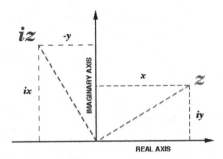

Figure 3.5 Geometric representation of multiplication by i. The point iz is obtained by 90° counterclockwise rotation of z in the complex plane.

Very often we need to transfer a factor i from a denominator to a numerator. The key result is

$$\frac{1}{i} = -i. \tag{3.79}$$

Several algebraic manipulations with complex numbers are summarized in the following equations:

$$(x_1 + iy_1) + (x_2 + iy_2) = (x_1 + x_2) + i(y_1 + y_2), \tag{3.80}$$

$$(x_1 + iy_1)(x_2 + iy_2) = (x_1 x_2) + i(x_1 y_2 + y_1 x_2) + i^2(y_1 y_2)$$
$$= (x_1 x_2 - y_1 y_2) + i(x_1 y_2 + y_1 x_2), \tag{3.81}$$

$$\frac{x_1 + iy_1}{x_2 + iy_2} = \frac{(x_1 + iy_1)(x_2 - iy_2)}{(x_2 + iy_2)(x_2 - iy_2)} = \frac{x_1 x_2 + y_1 y_2}{x_2^2 + y_2^2} + i\frac{x_2 y_1 - x_1 y_2}{x_2^2 + y_2^2}. \tag{3.82}$$

Note the strategy for expressing a fraction as a sum or real and imaginary parts: multiply by the complex conjugate of the denominator then recognize the square of an absolute value in the form of (3.75).

Problem 3.7.1. Find the real and imaginary parts of z^2, where $z = x + iy$, x and y real.

Problem 3.7.2. If you don't mind doing the algebra, analogously find the real and imaginary parts of z^3.

3.8 Factorials, Permutations and Combinations

Imagine that we have a dozen differently colored eggs which we need to arrange in an egg carton. The first egg can go into any one of 12 cups. The second egg can then go into any of the remaining 11 cups. So far, there are $12 \times 11 = 132$ possible arrangements for these two eggs in the carton. The third egg can go into 10 remaining cups. Continuing the placement of eggs, we will wind up with one of a possible total of $12 \times 11 \times 10 \times 9 \times 8 \times 7 \times 6 \times 5 \times 4 \times 3 \times 2 \times 1$ distinguishable arrangements. (This multiplies out to 479,001,600.) The product of a positive integer n with all the preceding integers down to 1 is called n factorial, designated $n!$:

$$n! \equiv n(n-1)(n-2)\cdots 1. \tag{3.83}$$

The first few factorials are $1! = 1, 2! = 2, 3! = 6, 4! = 24, 5! = 120$. As you can see, the factorial function increases rather steeply. Our original example involved $12! = 479,001,600$. The symbol for factorial is the same as an explanation point. (Thus be careful when you write something like, "To our amazement, the membership grew by 100!")

Can factorials also be defined for nonintegers? Later we will introduce the gamma function, which is a generalization of the factorial. Until then you can savor the amazing result that

$$\left(-\frac{1}{2}\right)! = \sqrt{\pi}. \tag{3.84}$$

Our first example established a fundamental result in combinational algebra: the number of ways of arranging n *distinguishable* objects (say, with different colors) in n different boxes equals $n!$. Stated another way, the number of possible *permutations* of n distinguishable objects equals $n!$.

Suppose we had started the preceding exercise with just a half-dozen eggs. The number of distinguishable arrangements in our egg carton would then be "only" $12 \times 11 \times 10 \times 9 \times 8 \times 7 = 665,280$. This is equivalent to $12!/6!$. The general result for the number of ways of permuting r distinguishable objects in n different boxes (with $r < n$) is given by

$$P(n, r) = \frac{n!}{(n-r)!}. \tag{3.85}$$

(You might also encounter the alternative notation $_nP_r$, ^nP_r, P_r^n, or $P_{n,r}$.) This formula also subsumes our earlier result, which can be written $P(n, n) = n!$. To be consistent with (3.85), we must interpret $0! = 1$.

Consider now an alternative scenario in which the eggs are not colored, so that they remain *indistinguishable*. The different results of our manipulations are now known as *combinations*. Carrying out an analogous procedure, our first egg can again go into 12 possible cups and the second into one of the remaining 11. But a big difference now is we can no longer tell which is the first egg and which is the second—remember they are indistinguishable. So the number of possibilities is reduced to $12 \times 11 \div 2$. After placing 3 eggs in $12 \times 11 \times 10$ available cups, the identical appearance of the eggs reduces the number of distinguishable arrangements by a factor of $3! = 6$. We should now be able to generalize for the number of combinations with m indistinguishable objects in n boxes. The result is

$$C(n, r) = \frac{P(n, r)}{r!} = \frac{n!}{(n-r)!r!}. \tag{3.86}$$

Another way of deducing this result. The total number of permutations of n objects is, as we have seen, equal to $n!$. Now, permutations can be of two types: *indistinguishable* and *distinguishable*. The eggs have $r!$ indistinguishable permutations among themselves, while the empty cups have $(n-r)!$ indistinguishable ways of hypothetically numbering them. Every other rearrangement is distinguishable. If $C(n,r)$ represents the total number of distinguishable configurations then

$$n! = C(n,r)r!(n-r)!, \tag{3.87}$$

which is equivalent to (3.86).

A simple generalization to distinguish permutations from combinations is that permutations are for lists, in which the order matters, while combinations are for groups in which the order doesn't matter.

Suppose you have n good friends seated around a dinner table who wish to toast one another by clinking wineglasses. How many "clinks" will you hear? The answer is the number of possible combinations of objects taken 2 at a time from a total of n, given by

$$C(n,2) = \frac{n!}{(n-2)!2!} = \frac{n(n-1)}{2}. \tag{3.88}$$

You can also deduce this more directly by the following argument. Each of n diners clinks wineglasses with his or her $n-1$ companions. You might first think there must be $n(n-1)$ clinks. But, if you listen carefully, you will realize that this counts each clink twice, one for each clinkee. Thus dividing by 2 gives the correct result $n(n-1)/2$.

Problem 3.8.1. A basketball team has 12 players. How many different ways can the coach choose the starting 5?

3.9 The Binomial Theorem

Let us begin with an exercise in experimental algebra:

$$
\begin{aligned}
(a+b)^0 &= 1 \\
(a+b)^1 &= a+b \\
(a+b)^2 &= a^2 + 2ab + b^2 \\
(a+b)^3 &= a^3 + 3a^2b + 3ab^2 + b^3 \\
(a+b)^4 &= a^4 + 4a^3b + 6a^2b^2 + 4ab^3 + b^4 \\
(a+b)^5 &= a^5 + 5a^4b + 10a^3b^2 + 10a^2b^3 + 5ab^4 + b^5 \\
(a+b)^6 &= a^6 + 6a^5b + 15a^4b^2 + 20a^3b^3 + 15a^2b^4 + 6ab^5 + b^6
\end{aligned}
$$
$$\cdots \quad \cdots \quad \cdots \tag{3.89}$$

The array of numerical coefficients in (3.89)

$$
\begin{array}{c}
1 \\
1 \quad 1 \\
1 \quad 2 \quad 1 \\
1 \quad 3 \quad 3 \quad 1 \\
1 \quad 4 \quad 6 \quad 4 \quad 1 \\
1 \quad 5 \quad 10 \quad 10 \quad 5 \quad 1 \\
1 \quad 6 \quad 15 \quad 20 \quad 15 \quad 6 \quad 1 \\
\cdots \quad \cdots \quad \cdots
\end{array}
\tag{3.90}
$$

is called *Pascal's triangle*. Note that every entry can be obtained by taking the sum of the two numbers diagonally above it, e.g. $15 = 5 + 10$. These numbers are called *binomial coefficients*. You can convince yourself that they are given by the same combinatorial formula as $C(n, r)$ in Eq. (3.86). The binomial coefficients are usually written $\binom{n}{r}$. Thus

$$
\binom{n}{r} = \frac{n!}{(n-r)!r!},
\tag{3.91}
$$

where each value of n, beginning with 0, determines a row in the Pascal triangle.

Setting $a = 1, b = x$, the *binomial formula* can be expressed

$$
(1+x)^n = \sum_{r=0}^{n-1} \binom{n}{r} x^r = 1 + nx + \frac{n(n-1)}{2!}x^2 + \frac{n(n-1)(n-2)}{3!}x^3 + \cdots .
\tag{3.92}
$$

This was first derived by Isaac Newton in 1666. Remarkably, the binomial formula is also valid for negative, fractional, and even complex values of n, which was proved by Niels Henrik Abel in 1826. (It is joked that Newton didn't prove the binomial theorem for noninteger n because he wasn't Abel.) Here are a few interesting binomial expansions which you can work out for yourself:

$$
(1+x)^{-1} = 1 - x + x^2 - x^3 + \cdots ,
\tag{3.93}
$$
$$
(1-x)^{-1} = 1 + x + x^2 + x^3 + \cdots ,
\tag{3.94}
$$
$$
\frac{1}{\sqrt{1-x}} = (1-x)^{-1/2} = 1 + \frac{1}{2}x + \frac{3}{8}x^2 + \frac{5}{16}x^3 + \frac{35}{128}x^4 + \cdots .
\tag{3.95}
$$

Each of the above series is convergent only for $|x| < 1$.

Problem 3.9.1. Determine the first few terms in the expansion of $\sqrt{1+x}$.

3.10 *e* is for Euler

Imagine there is a bank in your town that offers you 100% annual interest on your deposit (we will let pass the possibility that the bank might be engaged in questionable loan-sharking activities). This means that if you deposit $1 on January 1, you will get back $2 one year later. Another bank across town wants to get in on the action and offers 100% annual interest *compounded semiannually*. This means that you get 50% interest credited after half a year, so that your account is worth $1.50 on July 1. But this *total* amount then grows by another 50% in the second half of the year. This gets you, after 1 year,

$$\$(1 + 1/2)(1 + 1/2) = \$2.25. \tag{3.96}$$

A third bank picks up on the idea and offers to compound your money *quarterly*. Your $1 there would grow after a year to

$$\$(1 + 1/4)^4 = \$2.44. \tag{3.97}$$

Competition continues to drive banks to offer better and better compounding options, until the *Eulergenossenschaftsbank* apparently blows away all the competition by offering to compound your interest *continuously*—every second of every day! Let's calculate what your dollar would be worth there after 1 year. Generalization from Eq. (3.97) suggests that compounding n times a year produces $\$(1 + 1/n)^n$. Here are some numerical values for increasing n:

$$(1 + 1/5)^5 = 2.48832, \quad (1 + 1/10)^{10} = 2.59374, \quad (1 + 1/20)^{20} = 2.65330,$$
$$(1 + 1/50)^{50} = 2.69159, \quad (1 + 1/100)^{100} = 2.70481, \dots$$

$$\tag{3.98}$$

The ultimate result is

$$\lim_{n \to \infty} (1 + 1/n)^n = 2.718281828459 \dots \equiv e. \tag{3.99}$$

This number was designated e by the great Swiss mathematician Leonhard Euler (possibly after himself). Euler (pronounced approximately like "oiler") also first introduced the symbols i, π, and $f(x)$. After π itself, e is probably the most famous transcendental number, also with a never-ending decimal expansion. The tantalizing repetition of "1828" is just coincidental.

The binomial expansion applied to the expression for e gives

$$(1 + 1/n)^n = 1 + n\left(\frac{1}{n}\right) + \frac{n(n-1)}{2!}\left(\frac{1}{n}\right)^2 + \frac{n(n-1)(n-2)}{3!}\left(\frac{1}{n}\right)^3 + \cdots. \tag{3.100}$$

As $n \to \infty$, the factors $(n-1)$, $(n-2)$, ... all become insignificantly different from n. This suggests the infinite-series representation for e

$$e = \sum_{n=0}^{\infty} \frac{1}{n!} = 1 + 1 + \frac{1}{2!} + \frac{1}{3!} + \frac{1}{4!} + \cdots. \tag{3.101}$$

Remember that $0! = 1$ and $1! = 1$. This summation converges much more rapidly than the procedure of Eq. (3.99). After just six terms, we obtain the approximate value $e \approx 2.71667$.

Let's return to consideration of interest-bearing savings accounts, this time in more reputable banks. Suppose a bank offers $X\%$ annual interest. If $x = .01X$, your money would grows by a factor of $(1 + x)$ every year, without compounding. If we were able to get interest compounding n times a year, the net annual return would increase by a factor

$$(1 + x/n)^n = 1 + n\left(\frac{x}{n}\right) + \frac{n(n-1)}{2!}\left(\frac{x}{n}\right)^2 + \frac{n(n-1)(n-2)}{3!}\left(\frac{x}{n}\right)^3 + \cdots.$$

(3.102)

Note that, after defining $m = n/x$,

$$\lim_{n \to \infty} (1 + x/n)^n = \lim_{m \to \infty} (1 + 1/m)^{mx} = \left[\lim_{m \to \infty} (1 + 1/m)^m\right]^x = e^x.$$ (3.103)

Therefore, in the limit $n \to \infty$, Eq. (3.102) implies the series

$$e^x = \sum_{n=0}^{\infty} \frac{x^n}{n!} = 1 + x + \frac{x^2}{2!} + \frac{x^3}{3!} + \frac{x^4}{4!} + \cdots.$$ (3.104)

This defines the *exponential function*, which plays a major role in applied mathematics. The very steep growth of the factorials guarantees that the expansion will converge to a finite quantity for any finite value of x, real, imaginary, or complex. The inverse of the exponential function is the natural logarithm, defined in Eq. (3.55):

$$y = e^x \iff x = \ln y.$$ (3.105)

Two handy relations are

$$\ln(e^x) = x \quad \text{and} \quad e^{\ln x} = x.$$ (3.106)

When the exponent of the exponential is a complicated function, it is easier to write

$$e^X = \exp(X).$$ (3.107)

Exponential growth and exponential decay, sketched in Figure 3.6, are observed in a multitude of natural processes. For example, the population of a colony of bacteria, given unlimited nutrition, will grow exponentially in time:

$$N(t) = N_0 e^{kt},$$ (3.108)

where N_0 is the population at time $t = 0$ and k is a measure of the rate of growth. Conversely, a sample of a radioactive element will decay exponentially:

$$N(t) = N_0 e^{-kt}.$$ (3.109)

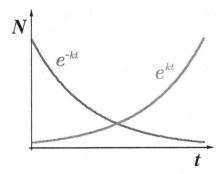

Figure 3.6 Exponential growth and decay.

A measure of the rate of radioactive decay is the *half-life* $t_{1/2}$, the time it takes for half of its atoms to disintegrate. The half-life can be related to the decay constant k by noting that after time $t = t_{1/2}$, N is reduced to $\frac{1}{2}N_0$. Therefore

$$\frac{1}{2} = e^{-kt_{1/2}} \tag{3.110}$$

and, after taking natural logarithms,

$$k = \frac{\ln 2}{t_{1/2}} \approx \frac{0.693}{t_{1/2}}. \tag{3.111}$$

No doubt, many of you will become fabulously rich in the future, due, in no small part, to the mathematical knowledge we are helping you acquire. You will probably want to keep a small fraction of your fortune in some CDs at your local bank. In better economic times there was a well-known "Rule of 72" which stated that to find the number of years required to double your principal at a given interest rate, just divide 72 by the interest rate. For example, at 8% interest, it would take about $72/8 \approx 9$ years. To derive this rule, assume that the principal $\$P$ will increase at an interest rate of $r\%$ to $\$2P$ in Y years, compounded annually. Thus

$$2P = P(1 + .01r)^Y. \tag{3.112}$$

Taking natural logarithms we can solve for

$$Y = \frac{\ln 2}{\ln(1 + .01r)}. \tag{3.113}$$

To get this into a neat approximate form $Y = \text{const}/r$, let

$$\frac{\ln 2}{\ln(1 + .01r)} \approx \frac{\text{const}}{r}. \tag{3.114}$$

You can then show for interest rates in the neighborhood of 8% ($r = 8$), the approximation will then work with the constant approximated by 72.

Problem 3.10.1.   Unfortunately, bank interest rates have lately fallen to the neighborhood of 1%. How many years would it take for you to double your money at this puny rate?

4 Trigonometry

4.1 What Use is Trigonometry?

Trigonometry, as its Greek and Latin roots suggest, is primarily the study of triangles. As we will see, circles also play a very prominent role. Trigonometric functions are, in fact, sometimes designated as *circular functions*. Figure 4.1 suggests how a clever caveman might determine the height y of a cliff by measuring the distance x to its base and the angle θ that he has to look upward. This could be done by making a scale drawing but, more elegantly, using trigonometry, the height y is equal to $x \tan \theta$.

4.2 Geometry of Triangles

In this section we review some useful elementary results for triangles, most of them dating back to Euclid's *Elements*. Figure 4.2 shows the essential elements of a triangle: the three sides, a, b, and c, the angles opposite them, A, B, and C, and the altitude h, perpendicular to the base b. There are also analogous altitudes perpendicular to a and c, not shown.

A signature feature of Euclidean space is that the angles of a triangle add up to exactly 180°:

$$A + B + C = 180°. \tag{4.1}$$

(By contrast, in spherical geometry, the sum of the angles is greater than 180°, while in hyperbolic geometry, it is less than 180°.)

The sides of a triangle always fulfill the *triangle inequalities*:

$$a + b > c, \quad a + c > b, \quad \text{and} \quad b + c > a. \tag{4.2}$$

Triangles can be classified according to the number of equal sides and angles, as shown on the top row of Figure 4.3. An equilateral triangle has three equal sides and three equal angles: $a = b = c$ and $A = B = C$. An isosceles triangle has two equal sides and two equal angles, for example: $a = b \neq c$ and $A = B \neq C$. Finally, a scalene triangle has three unequal sides and angles: $a \neq b \neq c$ and $A \neq B \neq C$. A right triangle has one angle equal to 90°. An acute triangle has all three of its angles less than 90°, while an obtuse triangle has one angle greater than 90°. The last three categories are illustrated in the bottom row of Figure 4.3.

Guide to Essential Math 2e. http://dx.doi.org/10.1016/B978-0-12-407163-6.00004-7

Figure 4.1 Stone-age trigonometry.

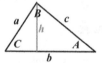

Figure 4.2 Sides, angles, and altitude of a triangle.

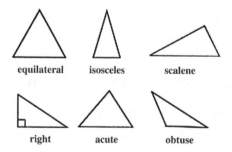

Figure 4.3 Classification of triangles.

The area of a triangle is equal to $\frac{1}{2}$ base $\times$ altitude. In terms of the variables shown in Figure 4.4,

$$A = \frac{1}{2}bh. \tag{4.3}$$

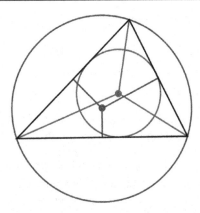

Figure 4.4 Incircle and circumcircle of triangle.

Another expression, known as Heron's (or Hero's) formula, computes the area in terms of its sides a, b, and c:

$$A = \sqrt{s(s-a)(s-b)(s-c)}, \tag{4.4}$$

where $s = (a + b + c)/2$, the semiperimeter of the triangle.

The bisectors of the three angle of a triangle intersect at a point known as the *incenter*, shown as a red[1] point in Figure 4.4. It is the center of the circle inscribed within the triangle, known as its *incircle*, also colored in red. Similarly, the perpendicular bisectors of the three sides meet at another point, called the *circumcenter*, which is the center of the *circumcircle*, which circumscribes the triangle. These latter features are shown in blue. Only for an equilateral triangle do the incenter and circumcenter coincide. The three altitudes of a triangle also meet at a point, which corresponds to the centroid (or center of gravity).

Problem 4.2.1.   The area of a trapezoid equals the product of its average width times its altitude:

$$A = \frac{(a+b)}{2}h.$$

Prove this by finding the sum of the areas of the red and blue triangles.

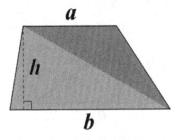

<hr>

[1] For interpretation of the references to color in this Figures 4.4 and 4.7 legend, the reader is referred to the web version of this book.

4.3   The Pythagorean Theorem

For a right triangle with sides a and b and hypotenuse c

$$a^2 + b^2 = c^2. \tag{4.5}$$

A pictorial proof of the theorem was given in Section 1.3. Albert Einstein, as a schoolboy, supposedly worked out his own proof of Pythagoras' theorem. His line of reasoning follows, although we may have changed the names of the variables he used. Figure 4.5 shows a right triangle cut by a perpendicular dropped to the hypotenuse from the opposite vertex. This produces three *similar* triangles since they all have the same angles α, β, and 90°. Each length a, b, and c represents the hypotenuse of one of the triangles. Since the area of each similar triangle is proportional to the square of its corresponding hypotenuse, we can write

$$E_a = ma^2,$$
$$E_b = mb^2,$$
$$E_c = mc^2. \tag{4.6}$$

The variables E might stand for area or extent (*Erstreckung* in German), while m is a proportionality constant (maybe *mengenproportional*). Since $E_c = E_a + E_b$, the ms cancel out and the result is Pythagoras' theorem (4.5).

The preceding story is perhaps an alternative interpretation of the famous Einstein cartoon reproduced in Figure 4.6.

Surprisingly, an alternative proof of the Pythagorean theorem was published by James A. Garfield, the 20th president of the United States. He is speculated to have been the most mathematically knowledgable US president. As shown in Figure 4.7, three right triangles are arranged to form a trapezoid (turned sideways from its usual orientation). The area of the trapezoid is equal to $(a + b)(a + b)/2$. The areas of the blue,[1] red, and purple triangles are, respectively, $ab/2$, $ba/2$, and $c^2/2$. Equating the two expressions gives

$$\frac{a^2}{2} + ab + \frac{b^2}{2} = ab + \frac{c^2}{2},$$

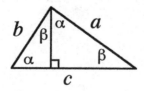

Figure 4.5 Einstein's proof of Pythagoras' theorem. The proof does assume, perhaps prematurely, that the angles of a triangle add up to 180°.

Figure 4.6 Einstein's proof of Pythagoras' theorem (?) ©2005 by Sidney Harris.

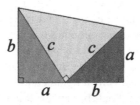

Figure 4.7 Garfield's proof of the Pythagorean theorem.

which reduces to $a^2 + b^2 = c^2$, giving Pythagoras' theorem for both the red and blue triangles.

The converse of Pythagoras' theorem is also valid. If three lengths a, b, and c satisfy Eq. (4.5), then they must form a right triangle with c as the hypotenuse. This provides a handy way for carpenters to construct a right angle: just mark off the lengths 3, 4, and 5 in., then the angle between the first two is equal to $90°$. There are, in fact, an infinite number of integer triples which satisfy $a^2 + b^2 = c^2$, beginning with $\{3,4,5\}$, $\{5,12,13\}$, $\{8,15,17\}$.

Analogous relations do *not* exist for powers of integers greater than 2. What has long been called "Fermat's last theorem" states that

$$x^n + y^n = z^n \tag{4.7}$$

has no nonzero integer solutions x, y, and z when $n > 2$. This is the best known instance of a *Diophantine equation*, which involves only integers. Fermat wrote in his notes around 1630, "I have discovered a truly remarkable proof which this margin is too small to contain." This turned out to have been a mischievous tease that took over three centuries to unravel. Some of the world's most famous mathematicians have since struggled with the problem. These efforts were not entirely wasted since they stimulated significant advances in several mathematical fields including analytic number theory and algebraic geometry. Fermat's *conjecture*, as it should have been called, was finally proven in 1993–1995 by the British mathematician Andrew Wiles, working at Princeton University. What we should now call the *Fermat-Wiles theorem*, took some 200 journal pages to present. (This whole book, let alone the margin, is too small to contain the proof!)

4.4 π in the Sky

The Babylonians (ca. 2400 BC) observed that the annual track of the Sun across the sky took approximately 360 days. Consequently, they divided its near-circular path into $360°$, as a measure of each day's progression. That is why we still count one spin around a circle as $360°$ and a right angle as $90°$. This way of measuring angles is not very fundamental from a mathematical point of view, however. Mathematicians prefer to measure distance around the circumference of a circle in units of the radius, defined as 1 radian (rad). The Greeks designated the ratio of the circumference to the diameter, which is twice the radius, as π. Thus $360°$ corresponds to 2π radians. A semicircle, $180°$, is π radians, while a right angle, $90°$, is $\pi/2$ radians. One radian equals approximately $57.3°$. You should be very careful to set your scientific calculators to "radians" when doing most trigonometric manipulations.

It has been known since antiquity that π is approximately equal to 3. The Old Testament (II Chronicles 4:2) contains a passage describing the building of Solomon's temple: "Also he made a molten sea of ten cubits from brim to brim, round in compass ... and a line of thirty cubits did compass it round about." This appears to imply that the ancient Hebrews used a value of $\pi \approx 3$. Two regular hexagons inscribed in and circumscribed around a circle (Figure 4.8) establish that the value of π lies in the range $3 < \pi < 3.46$.

As a practical matter, the value of π could be determined with a tape measure wound around a circular object. In carpentry and sewing, an adequate approximation is $\pi \approx 22/7 = 3\frac{1}{7}$. (In 1897, the Indiana House of Representatives decreed by a vote of 67–0 that π should be simplified to exactly 3.2. The measure however never reached the floor of the Indiana Senate.) Accurate computed values of π are usually obtained from power series for inverse trigonometric functions. Since we haven't introduced these yet, let us demonstrate a method for computing π which uses only the Pythagorean theorem. It was originally the idea of Archimedes to construct n-sided regular polygons inscribed in a unit circle (radius = 1). Then, as n becomes larger and larger, the perimeter of the polygon approaches the circumference of the circle. Denote the side of one such polygon by S_n. Then the perimeter equals nS_n and an estimate for π, which we can call π_n, is equal to $nS_n/2$. For example, the hexagon in Figure 4.8 gives $\pi_6 = 3$.

Figure 4.9 shows how to calculate the length of a side S_{2n} of a regular polygon with $2n$ sides from S_n, that of a polygon with n sides. Both polygons are inscribed in the unit circle with $OA = OB = OC = 1$. $AB = S_n$ is a side of the n-gon, while $AC = CB = S_{2n}$ are sides of the $2n$-gon. The segment $AD = S_n/2$ since the radius OC drawn to the new vertex C is a perpendicular bisector of side AB. Using the Pythagorean theorem, we can find two alternative expressions for the segment CD:

$$CD = \sqrt{AC^2 - AD^2} = \sqrt{S_{2n}^2 - (S_n/2)^2}$$

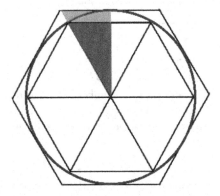

Figure 4.8 The perimeter of the inscribed hexagon equals three times the diameter of the circle. Noting that the two pink triangles are similar, you can show that the perimeter of the circumscribed hexagon equals $2\sqrt{3}$ times the diameter. (For interpretation of the references to color in this figure legend, the reader is referred to the web version of this article.)

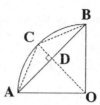

Figure 4.9 Detail of construction of an inscribed regular polygon with double the number of sides. $OA = OB = OC = 1$, the radius of a unit circle. AB represents a side S_n, while AC and CB are sides S_{2n}.

and

$$CD = 1 - OD = 1 - \sqrt{OA^2 - AD^2} = 1 - \sqrt{1 - (S_n/2)^2}.$$

Equating these, we obtain

$$S_{2n} = \sqrt{\frac{S_n^2}{4} + \left(1 - \sqrt{1 - \frac{S_n^2}{4}}\right)^2} = \sqrt{2 - \sqrt{4 - S_n^2}}, \qquad (4.8)$$

which relates the values of π_{2n} and π_n:

$$\pi_{2n} = n\sqrt{2 - \sqrt{4 - \left(\frac{2\pi_n}{n}\right)^2}}. \qquad (4.9)$$

Such an iterative evaluation was first carried out by Viète in 1593, starting with an inscribed square with $S_4 = \sqrt{2}$. This produced, for the first time, an actual formula for π, expressed as an infinite sequence of nested square roots:

$$\pi = \lim_{N \to \infty} 2^N \underbrace{\sqrt{2 - \sqrt{2 + \sqrt{2 + \cdots}}}}_{N}, \qquad (4.10)$$

where the underbrace N indicates the total number of square-root signs. (For compactness, $n - 1$ has been replaced by N.) For a $2^{10} = 1024$-sided polygon (corresponding to $N = 9$) this procedure gives a value accurate to six significant figures:

$$\pi = 3.14159\ldots \qquad (4.11)$$

A rational approximation giving the same accuracy is $\pi \approx 355/113$.

It has long been a popular macho sport to calculate π to more and more decimal places, using methods which converge much faster than the formulas we consider. The current (2012) record is held by a Japanese supercomputer, which gives π to over 10 trillion digits. While such exercises might have little or no practical value, they serve as tests for increasingly powerful supercomputers. Sequences of digits from π can be used to generate random numbers (technically *pseudorandom*) for use in Monte Carlo computations and other simulations requiring random input.

Those of you taking certain physics or chemistry courses might appreciate a cute mnemonic giving π to 15 digits (3.141 592 653 589 79). Adapted from James Jeans, it runs "Now I need a drink, alcoholic of course, after the heavy sessions involving quantum mechanics."

4.5 Sine and Cosine

Let us focus on one angle of a right triangle, designated by θ in Figure 4.10. We designate the two perpendicular sides as being *opposite* and *adjacent* to the angle θ. The sine and cosine are then defined as the ratios

$$\sin\theta \equiv \frac{\text{opposite}}{\text{hypotenuse}} \quad \text{and} \quad \cos\theta \equiv \frac{\text{adjacent}}{\text{hypotenuse}}. \tag{4.12}$$

Later we will also deal with the tangent:

$$\tan\theta \equiv \frac{\text{opposite}}{\text{adjacent}} = \frac{\sin\theta}{\cos\theta}. \tag{4.13}$$

A popular mnemonic for remembering which ratios go with which trigonometric functions is "SOHCAHTOA," which might be the name of your make-believe Native American guide through the trigonometric forest.

It is extremely instructive to represent the sine and cosine on the unit circle, shown in Figure 4.11, in which the hypotenuse corresponds to a radius equal to 1 unit. The circle is conveniently divided into four quadrants, I–IV, each with θ varying over an interval of $\pi/2$ radians from the range 0 to 2π. The lengths representing $\sin\theta$ are vertical lines on the unit circle while those representing $\cos\theta$ are horizontal. The cosine is closer to the angle—you might remember this by associating *c*osine with *c*ozy *up* and *s*ine with *stand off*. The functions shown in green have positive values, while those shown in red have negative values. Thus $\sin\theta$ is positive in quadrants I and II, negative in quadrants III and IV, while $\cos\theta$ is positive in quadrants I and IV, negative in quadrants II and III. It should also be clear from the diagram that, for real values of θ, $\sin\theta$ and $\cos\theta$ can have values only in the range $[-1, 1]$.

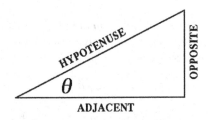

Figure 4.10 Right triangle used to define trigonometric functions.

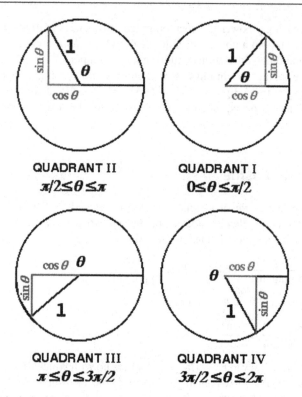

Figure 4.11 Unit circle showing $\sin \theta$ and $\cos \theta$ in each quadrant. Positive values of the functions are shown in green, negative, in red. (For interpretation of the references to color in this figure legend, the reader is referred to the web version of this article.)

Note that any angle θ less than 0 or greater than 2π is indistinguishable on the unit circle from one in the range 0 to 2π. The trigonometric functions are *periodic* in 2π and have the same values for any $\theta \pm 2n\pi$ with integer n. The functions $\sin \theta$ and $\cos \theta$ are plotted in Figure 4.12. These are commonly designated as *sinusoidal* functions or "sine waves." Note that sine and cosine have the same shape, being just displaced from one another by $\pi/2$. It can be seen that

$$\sin\left(\frac{\pi}{2} - \theta\right) = \cos\theta \quad \text{and} \quad \cos\left(\frac{\pi}{2} - \theta\right) = \sin\theta. \tag{4.14}$$

The angle $\frac{\pi}{2} - \theta$ is known as the *complement* of θ. Equation (4.14) is equivalent to a catchy-sounding rule: *The function of an angle is equal to the corresponding cofunction of its complement.* Note that cosine and sine are even and odd functions, respectively:

$$\cos(-\theta) = \cos\theta \quad \text{while} \quad \sin(-\theta) = -\sin\theta. \tag{4.15}$$

Whenever one of these functions goes through zero, the other has a local maximum at $+1$ or minimum at -1. This follows easily from differential calculus, as we will show later.

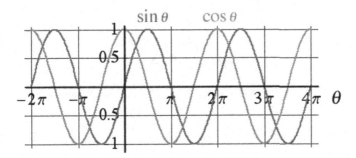

Figure 4.12 Plots of sine and cosine.

Pythagoras' theorem translates to the fundamental trigonometric identity:

$$\cos^2\theta + \sin^2\theta = 1. \tag{4.16}$$

Note that $(\cos\theta)^2$ and $(\sin\theta)^2$ are conventionally written $\cos^2\theta$ and $\sin^2\theta$. They are not to be confused, of course, with $\cos(\theta^2)$ and $\sin(\theta^2)$.

To compound the notational confusion, $\sin^{-1}x$ and $\cos^{-1}x$ are used to designate *inverse trigonometric functions*, also written arcsin x and arccos x, respectively. These inverse functions are related by the following correspondences:

$$x = \sin\theta \iff \theta = \arcsin x = \sin^{-1}x, \tag{4.17}$$
$$x = \cos\theta \iff \theta = \arccos x = \cos^{-1}x. \tag{4.18}$$

Since $\sin\theta$ and $\cos\theta$ are periodic functions, their inverse functions must be *multivalued*. For example, if arcsin $x = \theta$, it must likewise equal $\theta \pm 2n\pi$, for $n = 0, 1, 2, \ldots$ The *principal value* of arcsin x, sometimes designated Arcsin x, is limited to the range $[-\frac{\pi}{2}, \frac{\pi}{2}]$, corresponding to $-1 \leqslant x \leqslant +1$. Analogously, the principal value of arccos x, likewise designated Arccos x, lies in the range $[0, \pi]$. Graphs of arcsin x and arccos x can be obtained by turning Figure 4.12 counterclockwise by $90°$ and then reflecting in the θ-axis.

Several values of sine and cosine occur so frequently that they are worth remembering. Almost too obvious to mention,

$$\sin 0 = \cos\left(\frac{\pi}{2}\right) = 0, \quad \cos 0 = \sin\left(\frac{\pi}{2}\right) = 1. \tag{4.19}$$

When $\theta = \pi/4$ or $45°$, $\cos\theta = \sin\theta$, so that (4.16) implies

$$\cos\left(\frac{\pi}{4}\right) = \sin\left(\frac{\pi}{4}\right) = \frac{1}{\sqrt{2}} = \frac{\sqrt{2}}{2} \approx 0.707, \tag{4.20}$$

which is a factor well known in electrical engineering, in connection with the rms voltage and current of an AC circuit. An equilateral triangle with side 1, cut in half, gives a "30–60–90 triangle." Since $30° = \pi/6$ and $60° = \pi/3$, you can show using

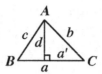

Figure 4.13 Diagram for law of sines and law of cosines.

Pythagoras' theorem that

$$\sin\left(\frac{\pi}{6}\right) = \cos\left(\frac{\pi}{3}\right) = \frac{1}{2}, \quad \sin\left(\frac{\pi}{3}\right) = \cos\left(\frac{\pi}{6}\right) = \frac{\sqrt{3}}{2}. \tag{4.21}$$

As $\theta \to 0$, in the first quadrant of Figure 4.11, the length of the line representing $\sin\theta$ approaches the magnitude of the arc of angle θ—remember this is measured in radians. This implies a very useful approximation

$$\sin\theta \approx \theta \quad \text{for } \theta \to 0. \tag{4.22}$$

For a triangle of arbitrary shape, not limited to a right triangle, two important relations connecting the lengths a,b,c of the three sides with the magnitudes of their opposite angles A, B, C can be derived. In Figure 4.13, a perpendicular is drawn from any side to its opposite angle, say from angle A to side a. Call this length d. It can be seen that $d = c\sin B$ and also that $d = b\sin C$. Analogous relations can be found involving a and A. The result is the *law of sines*:

$$\frac{\sin A}{a} = \frac{\sin B}{b} = \frac{\sin C}{c}. \tag{4.23}$$

The little triangle to the right of line d has a base given by $a' = b\cos C$. Therefore the base of the little triangle to the left of d equals $a - a' = a - b\cos C$. Using $d = b\sin C$ again, and applying Pythagoras' theorem to the left-hand right triangle, we find

$$c^2 = d^2 + (a - a')^2 = b^2\sin^2 C + a^2 - 2ab\cos C + b^2\cos^2 C. \tag{4.24}$$

The identity (4.16) simplifies the equation, leading to the *law of cosines*:

$$c^2 = a^2 + b^2 - 2ab\cos C \quad et \; cyc. \tag{4.25}$$

By *et cyc* we mean that the result holds for all cyclic permutations $\{a, A \to b, B \to c, C \to a, A\}$. The law of cosines is clearly a generalization of Pythagoras' theorem, valid for *all* triangles. It reduces to Pythagoras' theorem when $C = \pi/2$, so that $\cos C = 0$.

4.6   Tangent and Secant

Additional subsidiary trigonometric functions can be defined in terms of sine and cosine. Consider just the first quadrant of the unit circle, redrawn in Figure 4.14.

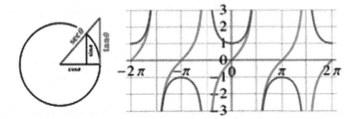

Figure 4.14 Graphs of tangent (red) and secant (blue). (For interpretation of the references to color in this figure legend, the reader is referred to the web version of this article.)

Let the horizontal line containing $\cos\theta$, as well as the hypotenuse, be extended until they intersect the circle. A vertical tangent line of length $\tan\theta$ can then be used to construct a larger right triangle. The new hypotenuse is called a *secant* and labeled $\sec\theta$. Clearly the new triangle is similar to the original one, with its horizontal side being equal to 1, the unit-circle radius. Therefore

$$\frac{\tan\theta}{1} = \frac{\sin\theta}{\cos\theta}. \tag{4.26}$$

Also by Pythagoras' theorem,

$$\sec^2\theta - \tan^2\theta = 1, \tag{4.27}$$

so that

$$\sec\theta = \sqrt{1+\tan^2\theta} = \sqrt{\frac{\cos^2\theta + \sin^2\theta}{\cos^2\theta}} = \frac{1}{\cos\theta} \tag{4.28}$$

making use of Eq. (4.16). The tangent and secant are plotted in Figure 4.14. Since $|\cos\theta| \leqslant 1$, we find $|\sec\theta| \geqslant 1$, since the secant is the reciprocal of the cosine. Note that $\tan\theta$ is periodic in π, rather than 2π, and can take any value from $-\infty$ to $+\infty$.

It is also possible to define cofunctions in analogy with (4.14), namely the cotangent and cosecant:

$$\tan\left(\frac{\pi}{2} - \theta\right) \equiv \cot\theta \quad \text{and} \quad \sec\left(\frac{\pi}{2} - \theta\right) \equiv \csc\theta. \tag{4.29}$$

This completes the list of standard trigonometric functions. In terms of sine and cosine:

$$\tan\theta = \frac{\sin\theta}{\cos\theta}, \quad \cot\theta = \frac{1}{\tan\theta} = \frac{\cos\theta}{\sin\theta}, \quad \sec\theta = \frac{1}{\cos\theta}, \quad \csc\theta = \frac{1}{\sin\theta}. \tag{4.30}$$

4.7 Trigonometry in the Complex Plane

In Figure 3.4, a complex quantity z was represented by its Cartesian coordinates x and y. Alternatively, a point in the x-, y-plane can be represented in *polar coordinates*, usually designated r and θ. Conventionally, r is the distance from the origin, while θ is the angle that the vector $\mathbf{r}$ makes with the positive x-axis. The coordinates in the two systems are related by

$$x = r\cos\theta, \quad y = r\sin\theta \quad \text{or} \quad r = \sqrt{x^2 + y^2}, \quad \theta = \arctan\left(\frac{y}{x}\right). \tag{4.31}$$

A complex quantity $z = x + iy$ expressed in polar coordinates is called a *phasor*. In place of r we have the modulus $|z|$. The angle θ, now called the *phase* or the *argument*, plays the same role as in Figure 4.11, with the unit circle generalized to a circle of radius $|z|$. Thus the phasor representation of a complex number takes the form

$$z = |z|(\cos\theta + i\sin\theta). \tag{4.32}$$

A convenient abbreviation which we will employ for a short while is

$$\text{cis}\,\theta \equiv \cos\theta + i\sin\theta. \tag{4.33}$$

In Figure 3.5 it was shown how multiplication of a complex numbers by i could be represented by a 90° rotation in the complex plane. Note that i can be represented by the phasor

$$i = \text{cis}\left(\frac{\pi}{2}\right), \tag{4.34}$$

while

$$iz = |z|\text{cis}\left(\theta + \frac{\pi}{2}\right). \tag{4.35}$$

We can generalize that the product of two arbitrary phasors, say $z_1 = |z_1|\text{cis}\,\theta_1$ and $z_2 = |z_2|\text{cis}\,\theta_2$, is given by

$$z_1 z_2 = |z_1||z_2|\text{cis}(\theta_1 + \theta_2) \tag{4.36}$$

obtained by *multiplying* the two moduli while *adding* the two phases, as shown in Figure 4.15.

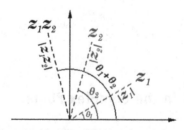

Figure 4.15 Multiplication of phasors.

Consider the multiplication of the two phasors $z_1 = \text{cis}\,\alpha$ and $z_2 = \text{cis}\,\beta$ to give $z_1 z_2 = \text{cis}(\alpha + \beta)$. Written out in full, this gives

$$
\begin{aligned}
(\cos\alpha + i\sin\alpha)(\cos\beta + i\sin\beta) &= \cos(\alpha + \beta) + i\sin(\alpha + \beta) \\
&= (\cos\alpha\cos\beta - \sin\alpha\sin\beta) \\
&\quad + i(\sin\alpha\cos\beta + \cos\alpha\sin\beta). \qquad (4.37)
\end{aligned}
$$

Equating the separate real and imaginary parts of each side of the equation, we obtain the two fundamental angle-sum trigonometric identities:

$$
\sin(\alpha + \beta) = \sin\alpha\cos\beta + \cos\alpha\sin\beta \qquad (4.38)
$$

and

$$
\cos(\alpha + \beta) = \cos\alpha\cos\beta - \sin\alpha\sin\beta. \qquad (4.39)
$$

From (4.38) and (4.39) we can derive the corresponding relation for $\tan(\alpha + \beta)$:

$$
\tan(\alpha + \beta) = \frac{\sin(\alpha + \beta)}{\cos(\alpha + \beta)} = \frac{\sin\alpha\cos\beta + \cos\alpha\sin\beta}{\cos\alpha\cos\beta - \sin\alpha\sin\beta}. \qquad (4.40)
$$

Dividing both numerator and denominator by $\cos\alpha\cos\beta$ and introducing $\tan\alpha$ and $\tan\beta$, we obtain

$$
\tan(\alpha + \beta) = \frac{\tan\alpha + \tan\beta}{1 - \tan\alpha\tan\beta}. \qquad (4.41)
$$

Problem 4.7.1.   As a special case of Eqs. (4.38) and (4.39), derive formulas for $\sin(2\theta)$ and $\cos(2\theta)$:

$$
\sin(2\theta) = 2\sin\theta\cos\theta, \qquad (4.42)
$$
$$
\cos(2\theta) = \cos^2\theta - \sin^2\theta. \qquad (4.43)
$$

Problem 4.7.2.   Using the substitution $\theta \to \theta/2$, derive the half-angle formulas:

$$
\sin\left(\frac{\theta}{2}\right) = \sqrt{\frac{1 - \cos\theta}{2}}, \qquad (4.44)
$$

$$
\cos\left(\frac{\theta}{2}\right) = \sqrt{\frac{1 + \cos\theta}{2}}. \qquad (4.45)
$$

4.8   de Moivre's Theorem

Equation (4.36) can be applied to the square of a phasor $z = \text{cis}\,\theta$, of modulus 1, giving $z^2 = (\text{cis}\,\theta)^2 = \text{cis}(2\theta)$. This can, in fact, be extended to the nth power of z giving

$(\operatorname{cis}\theta)^n = \operatorname{cis}(n\theta)$. This is a famous result known as *de Moivre's theorem*, which we write out in full:

$$(\cos\theta + i\sin\theta)^n = \cos(n\theta) + i\sin(n\theta). \tag{4.46}$$

Beginning with de Moivre's theorem, useful identities involving sines and cosines can be derived. For example, setting $n = 2$,

$$(\cos\theta + i\sin\theta)^2 = \cos^2\theta - \sin^2\theta + 2i\cos\theta\sin\theta = \cos(2\theta) + i\sin(2\theta). \tag{4.47}$$

Equating the real and imaginary parts on each side of the equation, we obtain the two identities

$$\sin(2\theta) = 2\sin\theta\cos\theta \tag{4.48}$$

and

$$\cos(2\theta) = \cos^2\theta - \sin^2\theta. \tag{4.49}$$

Analogously, with $n = 3$ in (4.46) we can derive:

$$\sin(3\theta) = 3\sin\theta - 4\sin^3\theta, \quad \cos(3\theta) = 4\cos^3\theta - 3\cos\theta. \tag{4.50}$$

Problem 4.8.1. Verify the expressions for $\sin(3\theta)$ and $\cos(3\theta)$.

de Moivre's theorem can be used to determine the nth roots of unity, namely the n complex roots of the equation

$$z^n = 1. \tag{4.51}$$

Setting $\theta = 2\pi k/n$ with $k = 0, 1, 2, \ldots, n-1$ in Eq. (4.46), we find

$$\left[\cos\left(\frac{2k\pi}{n}\right) + i\sin\left(\frac{2k\pi}{n}\right)\right]^n = \cos(2k\pi) + i\sin(2k\pi). \tag{4.52}$$

But, for integer k, $\cos(2k\pi) = 1$ while $\sin(2k\pi) = 0$. Thus the nth roots of unity are given by

$$z_{k,n} = \cos\left(\frac{2k\pi}{n}\right) + i\sin\left(\frac{2k\pi}{n}\right), \quad k = 0, 1, 2, \ldots, n-1 \tag{4.53}$$

When the nth roots of unity are plotted on the complex plane, they form a regular polygon with n sides, with one vertex at 1. For example, for $n = 2$, $z = \{+1, -1\}$, for $n = 3$, $z = \{+1, -\frac{1}{2} + i\frac{\sqrt{3}}{2}, -\frac{1}{2} - i\frac{\sqrt{3}}{2}\}$, for $n = 4$, $z = \{+1, +i, -1, -i\}$. The sum of the roots for each n adds to zero.

4.9 Euler's Theorem

de Moivre's theorem, Eq. (4.46), remains valid even for noninteger values of n. Replacing n by $1/m$ we can write

$$(\cos\theta + i\sin\theta)^{1/m} = \cos\left(\frac{\theta}{m}\right) + i\sin\left(\frac{\theta}{m}\right) \qquad (4.54)$$

or

$$(\cos\theta + i\sin\theta) = \left[\cos\left(\frac{\theta}{m}\right) + i\sin\left(\frac{\theta}{m}\right)\right]^m. \qquad (4.55)$$

In the limit as $m \to \infty$, $\cos(\theta/m) \approx \cos 0 = 1$. At the same time $\sin(\theta/m) \approx \theta/m$, as noted in Eq. (4.22). We can therefore write

$$(\cos\theta + i\sin\theta) = \lim_{m\to\infty}\left(1 + \frac{i\theta}{m}\right)^m. \qquad (4.56)$$

The limit defines the exponential function, as shown in Eq. (3.103). We arrive thereby at a truly amazing relationship:

$$e^{i\theta} = \cos\theta + i\sin\theta \qquad (4.57)$$

known as *Euler's theorem*. (This is actually one of at least 13 theorems, formulas, and equations which goes by this name. Euler was very prolific!)
  A very notable special case of Eq. (4.57), for $\theta = \pi$, is

$$e^{i\pi} = -1 \qquad (4.58)$$

an unexpectedly simple connection between the three mathematical entities, e, i, and π, each of which took us an entire section to introduce. This result can also be rearranged to

$$e^{i\pi} + 1 = 0 \qquad (4.59)$$

sometimes called *Euler's identity*. Several authors regard this as the most beautiful equation in all of mathematics. It contains what are perhaps the five most fundamental mathematical quantities: in addition to e, i, and π, the additive identity 0 and the multiplicative identity 1. It also makes use of the concepts of addition, multiplication, exponentiation, and equality. Because it represents so much in one small package, the formula has been imprinted on the side of some far-ranging NASA spacecraft to demonstrate the existence of intelligent life on Earth. It is to be hoped, of course, that extraterrestrials would be able to figure out that those symbols represent a mathematical equation and not a threat of interstellar war.
  Solving (4.57) and its complex-conjugate equation $e^{-i\theta} = \cos\theta - i\sin\theta$ for $\sin\theta$ and $\cos\theta$, we can represent these trigonometric functions in terms of complex exponentials:

$$\sin\theta = \frac{e^{i\theta} - e^{-i\theta}}{2i} \qquad (4.60)$$

and

$$\cos\theta = \frac{e^{i\theta} + e^{-i\theta}}{2}.$$

(4.61)

The power-series expansion for the exponential function given in Eq. (3.104) is, as we have noted, valid even for imaginary values of the exponent. Replacing x by $i\theta$ we obtain

$$e^{i\theta} = 1 + i\theta + \frac{(i\theta)^2}{2!} + \frac{(i\theta)^3}{3!} + \frac{(i\theta)^4}{4!} + \frac{(i\theta)^5}{5!} + \cdots.$$

(4.62)

Successive powers of i are given sequentially by

$$i^2 = -1, \ i^3 = -i, \ i^4 = 1, \ i^5 = i, \ \ldots$$

(4.63)

Collecting the real and imaginary parts of (4.62) and comparing with the corresponding terms in Euler's theorem (4.57) result in power-series expansions for the sine and cosine:

$$\sin\theta = \theta - \frac{\theta^3}{3!} + \frac{\theta^5}{5!} - \cdots$$

(4.64)

and

$$\cos\theta = 1 - \frac{\theta^2}{2!} + \frac{\theta^4}{4!} - \cdots.$$

(4.65)

4.10 Hyperbolic Functions

Hyperbolic functions are "copy-cats" of the corresponding trigonometric functions, in which the complex exponentials in Eqs. (4.60) and (4.61) are replaced by real exponential functions. The *hyperbolic sine* and *hyperbolic cosine* are defined, respectively, by

$$\sinh x \equiv \frac{e^x - e^{-x}}{2} \quad \text{and} \quad \cosh x \equiv \frac{e^x + e^{-x}}{2}.$$

(4.66)

Actually, hyperbolic functions result when sine and cosine are given imaginary arguments. Thus

$$\sin(ix) = i \sinh x \quad \text{and} \quad \cos(ix) = \cosh x.$$

(4.67)

The hyperbolic sine and cosine functions are plotted in Figure 4.16. Unlike their trigonometric analogs, they are not periodic functions and both have the domains $-\infty \leqslant x \leqslant \infty$. Note that as $x \to \infty$ both $\sinh x$ and $\cosh x$ approach $e^x/2$. The hyperbolic cosine represents the shape of a flexible wire or chain hanging from two fixed points, called a *catenary* (from the Latin *catena* = chain).

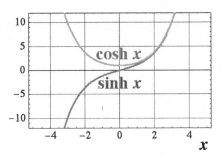

Figure 4.16 Hyperbolic sine and cosine.

By solving Eq. (4.66) for e^x and e^{-x}, we obtain the analog of Euler's theorem for hyperbolic functions:

$$e^{\pm x} = \cosh x \pm \sinh x. \tag{4.68}$$

The identity $e^x e^{-x} = 1$ then leads to the hyperbolic analog of (4.16):

$$\cosh^2 x - \sinh^2 x = 1. \tag{4.69}$$

The trigonometric sine and cosine are called *circular functions* because of their geometrical representation using the unit circle $x^2 + y^2 = 1$. The *hyperbolic functions* can analogously be based on the geometry of the *unit hyperbola $x^2 - y^2 = 1$*. We will develop the properties of hyperbolas, and other conic sections, in more detail in the following chapter. It will suffice for now to show the analogy with circular functions. Figure 4.17 shows the first quadrant of the unit circle and the unit hyperbola, each with specific areas A_c and A_h, respectively, shown shaded. For the circle, the area is equal to the fraction $\theta/2\pi$ of π, the area of the unit circle. Thus

$$\theta = 2A_c. \tag{4.70}$$

For the unit hyperbola, we will be able to compute the area $A_h = \frac{1}{2}\ln(x + y)$ later using calculus. It will suffice for now to define a variable

$$t = 2A_h. \tag{4.71}$$

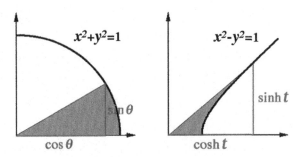

Figure 4.17 Geometric representation of circular and hyperbolic functions. The argument of each function equals twice the corresponding shaded area.

Analogous constructions in Figure 4.17 can then be used to represent the trigonometric functions $\sin\theta$, $\cos\theta$ and the hyperbolic functions $\sinh t$, $\cosh t$.

The series expansions for the hyperbolic functions are similar to (4.64) and (4.65), except that all terms have plus signs:

$$\sinh x = x + \frac{x^3}{3!} + \frac{x^5}{5!} + \cdots \tag{4.72}$$

and

$$\cosh x = 1 + \frac{x^2}{2!} + \frac{x^4}{4!} + \cdots . \tag{4.73}$$

Problem 4.10.1. Derive the analogs of Eqs. (4.38) and (4.39) for hyperbolic sine and cosine.

5 Analytic Geometry

The great French mathematician and philosopher René Descartes (1596–1650) is usually credited with developing the network of relationships that exist between algebraic quantities—such as numbers and functions—and their geometrical analogs—points and curves. Descartes was also responsible for the notion that the world is made up of two fundamentally different substances, mind and matter. From our enlightened modern viewpoint, Cartesian mind/body dualism is almost certainly misguided, but that does not compromise Descartes' contributions when he stuck to mathematics. (Descartes is also responsible for the intellectual's *raison d'être*, "*cogito ergo sum*"—I think therefore I am.)

5.1 Functions and Graphs

A function in mathematics is a relation that associates corresponding members of two different sets. For example, the expression

$$y = f(x) \tag{5.1}$$

says that if you give me a quantity x, called the *independent* variable, I have a rule which will produce a quantity y, the *dependent* variable. Often such a functional relation is written more compactly as $y(x)$. A function is called *single-valued* if x uniquely determines y, such as $y = x^2$ or $y = \sin x$, but multivalued if x can give more than one possible value of y. For example, $y = \sqrt{x}$ is double-valued since it can equal $\pm|\sqrt{x}|$ and $y = \arcsin x$ has an infinite number of values, $\arcsin x \pm 2n\pi$. The set of values x for which a function is defined is known as its *domain*. The domain might be limited in specific cases, say, to real x or to $0 \leqslant x \leqslant 1$. The possible values of y consistent with the choice of domain are known as its *range*. For example, $\sin x$ and $\cos x$ have the range $[-1, +1]$ when the domain of x is the real numbers.

If dependent and independent variables are not explicitly distinguished, their relationship can be expressed as an *implicit function*

$$f(x, y) = 0. \tag{5.2}$$

For example, $f(x, y) = x^2 + y^2 - 1 = 0$, the equation for the unit circle, can be solved for either $x = \sqrt{1 - y^2}$ or $y = \sqrt{1 - x^2}$.

In this chapter, we will be considering only relations involving two variables. Later we will generalize to more variables. A functional relation $f(x, y) = 0$ can be represented by a curve on the two-dimensional x, y-plane, a *Cartesian* coordinate system.

Guide to Essential Math 2e. http://dx.doi.org/10.1016/B978-0-12-407163-6.00005-9

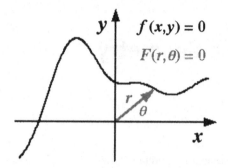

Figure 5.1 Plot of functional relation $f(x, y) = 0$ in Cartesian coordinates or the corresponding relation $F(r, \theta) = 0$ in polar coordinates.

The distance between two points (x_1, y_1) and (x_2, y_2) can be found using Pythagoras' theorem:

$$d_{12} = \sqrt{(x_2 - x_1)^2 + (y_2 - y_1)^2}. \tag{5.3}$$

We will also be using *polar coordinates*, in which a point on the plane is represented by its distance from the origin r, and its direction with respect to the x-axis, designated by the polar angle θ. The Cartesian and polar coordinate systems are related by

$$x = r\cos\theta, \quad y = r\sin\theta \quad \text{or} \quad r = \sqrt{x^2 + y^2}, \quad \theta = \arctan\left(\frac{y}{x}\right), \tag{5.4}$$

which we have already encountered in the phasor representation of complex numbers. Figure 5.1 shows schematically how a functional relation, $f(x, y) = 0$, or its polar equivalent $F(r, \theta) = 0$, can be represented in these alternative coordinate systems.

   As certain functions become familiar to you, association with the shape of their curves will become almost reflexive. As Frank Lloyd Wright might have put it, "form follows function."

5.2   Linear Functions

The simplest functional relation between two variables has the general form

$$Ax + By + C = 0, \tag{5.5}$$

which can represented by a straight line. The standard form in Cartesian coordinates is

$$y = mx + b, \tag{5.6}$$

which is sketched in Figure 5.2. It is to your future advantage to be able to understand linear equations from several different points of view. A linear relation means that increasing or decreasing the independent variable x by a given amount will cause

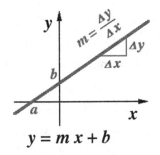

$$y = m x + b$$

Figure 5.2 Standard form for equation of straight line. a, b, and m are the x-intercept, y-intercept and slope, respectively. The alternative intercept form of the equation can be written $\frac{x}{a} + \frac{y}{b} = 1$.

a proportionate increase or decrease in the dependent variable y. The *y-intercept* is the point where the line crosses the y-axis, where $x = 0$. Equation (5.6) shows that occurs at $y = b$. The x-intercept, where $y = 0$, occurs at the point $x = -b/m \equiv a$. An alternative form for the equation in terms of its intercepts is

$$\frac{x}{a} + \frac{y}{b} = 1. \tag{5.7}$$

The parameter m in Eq. (5.6) is called the *slope*. It measures how steeply y rises or falls with x. Differential calculus, which we begin in the next chapter, is at its most rudimentary level, the computation of slopes of functions at different points. The slope tells how many units of y you go up or down when you travel along one unit of x. Symbolically,

$$m = \frac{\Delta y}{\Delta x}, \tag{5.8}$$

where Δ in universal mathematical usage represents the *change* in the quantity to which it is affixed. For example,

$$\Delta x = x_2 - x_1, \quad \Delta y = y_2 - y_1. \tag{5.9}$$

Slope might be used to describe the degree of inclination of a road. If a mountain road rises 5 m for every 100 m of horizontal distance on the map (which would be called a "5% grade"), the slope equals $5/100 = 0.05$. In the following chapter on differential calculus, we will identify the slope with the first derivative, using the notation:

$$m = \frac{dy}{dx}. \tag{5.10}$$

Only for a straight line does $dy/dx = \Delta y/\Delta x$. For other functions, we will need to consider the limits as Δx and $\Delta y \to 0$.

Note that $\Delta y/\Delta x$ has the form of a tangent of a right triangle with sides Δy and Δx (opposite/adjacent). If θ is the angle that the line makes with the horizontal (x) axis, we can identify

$$m = \tan \theta. \tag{5.11}$$

A horizontal line, representing an equation $y = $ const, has slope $m = 0$. A vertical line, representing an equation $x = $ const, has slope $m = \infty$. Another line drawn perpendicular to the given straight line with slope m must make an angle $\theta - \pi/2$ with the horizontal. The slope of the perpendicular, say $m_\perp$, is then given by

$$m_\perp = \tan\left(\theta - \frac{\pi}{2}\right) = -\tan\left(\frac{\pi}{2} - \theta\right) = -\cot\theta = -\frac{1}{\tan\theta} = -\frac{1}{m} \tag{5.12}$$

having exploited several trigonometric relations. We arrive thereby at a general result for the slopes of two mutually perpendicular curves at their point of intersection:

$$mm_\perp = -1. \tag{5.13}$$

One of Euclid's axioms was that through any two points, one and only one straight line can be drawn. Translated into analytic geometry this implies that two points (x_1, y_1) and (x_2, y_2) suffice to determine the equation for a straight line. In principle, the equation can be determined by writing Eq. (5.6) successively using (x_1, y_1) and (x_2, y_2) and solving the simultaneous equations for m and b; or, alternatively, solving Eq. (5.7) for a and b. A more illuminating approach is to focus on the slope of the line. Since the slope of a straight line is constant, we can write (5.8) using Δx and Δy computed between *any* two points. For example between the points (x_1, y_1) and (x_2, y_2),

$$m = \frac{y_2 - y_1}{x_2 - x_1}, \tag{5.14}$$

which determines the slope m. If (x, y) stands for some arbitrary point on the line then, likewise,

$$m = \frac{y - y_1}{x - x_1}. \tag{5.15}$$

This equation can be solved for y to give an equation for the line

$$y = y_1 + m(x - x_1). \tag{5.16}$$

With the identification $b = y_1 - mx_1$, this takes the standard form of Eq. (5.6).

Problem 5.2.1. Determine the equation of the straight line that passes through the points $(-1, 1)$ and $(1, -1)$.

Problem 5.2.2. Derive an expression for the distance from the point (x_0, y_0) to the line $Ax + By + C = 0$.

5.3 Conic Sections

Mathematically, a right-circular cone is a surface swept out by a straight line, with one point—the vertex—kept fixed while the line sweeps around a circular path. You would have to place *two* ice-cream cones point-to-point to simulate a mathematical cone, as

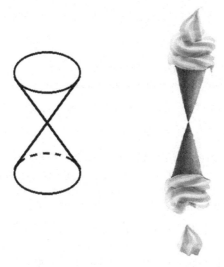

Figure 5.3 Mathematical definition of a cone and one possible physical realization. Each of the two sheets on opposite sides of the vertex is called a *nappe*.

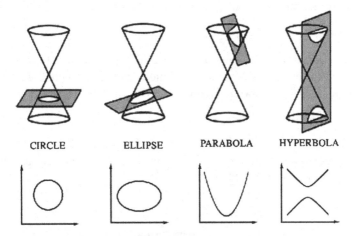

Figure 5.4 Intersections of the cone and plane showing how the different conic sections are generated.

shown in Figure 5.3. Greek mathematicians (Apollonius is usually credited) discovered that planes intersecting the cone at different angles produce several interesting curves, which are called *conic sections*. Degenerate cases, in which the plane passes through the vertex, give either a single point, a straight line, or two intersecting lines. The nondegenerate conic sections, illustrated in Figure 5.4, are circles, ellipses, parabolas, and hyperbolas.

A right-circular cone can be represented by the three-dimensional equation:

$$x^2 + y^2 = z^2. \tag{5.17}$$

For each value $\pm z$, this corresponds to a circle of radius z. Also, by analogy with Eq. (5.7) for a straight line, we surmise that the equation for a plane in three dimensions has the form

$$\frac{x}{a} + \frac{y}{b} + \frac{z}{c} = 1. \tag{5.18}$$

The intersection of the cone with an arbitrary plane is satisfied by points (x, y, z) which are simultaneous solutions of Eqs. (5.17) and (5.18). The equation for the surface of intersection, in the form $f(x, y) = 0$, can be obtained by eliminating z between the two equations. This implies that a conic section has the general form

$$Ax^2 + Bxy + Cy^2 + Dx + Ey + F = 0. \tag{5.19}$$

Assuming it is nondegenerate, the form of the conic section is determined by its *discriminant*, $B^2 - 4AC$. If the discriminant is positive, it is a hyperbola, if it is negative, an ellipse or a circle. A discriminant of zero implies a parabola. The coefficients D, E, and F help determine the location and scale of the figure.

The simplest nondegenerate conic section is the circle. A circle centered at (x_0, y_0) with a radius of a satisfies the equation

$$(x - x_0)^2 + (y - y_0)^2 = a^2. \tag{5.20}$$

The unit circle is the special case when $x^2 + y^2 = 1$. A circle viewed from an angle has the apparent shape of an ellipse. An ellipse centered at (x_0, y_0) with semimajor axis a and semiminor axis b, as shown in Figure 5.5, is described by the equation

$$\frac{(x - x_0)^2}{a^2} + \frac{(y - y_0)^2}{b^2} = 1. \tag{5.21}$$

It is assumed that $a > b$, so that the long axis of the ellipse is oriented in the x-direction. When $b = a$, the ellipse degenerates into a circle.

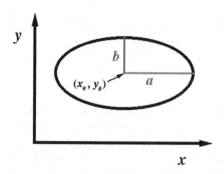

Figure 5.5 Ellipse with semimajor axis a and semiminor axis b.

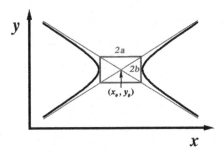

Figure 5.6 Hyperbola with semimajor axis a and semiminor axis b. Asymptotes are shown in gray.

Figure 5.6 represents a hyperbola with the equation:

$$\frac{(x - x_0)^2}{a^2} - \frac{(y - y_0)^2}{b^2} = 1. \tag{5.22}$$

The hyperbola can likewise be characterized by a semimajor axis a and a semiminor axis b, which, in this case, define a rectangle centered about (x_0, y_0). The two *branches* of the hyperbola are tangent to the rectangle, as shown. A distinctive feature are the *asymptotes*, the two diagonals of the rectangle extended to infinity. Their equations are

$$(y - y_0) = \pm\sqrt{\frac{b}{a}}(x - x_0). \tag{5.23}$$

When $b = a$, the asymptotes become perpendicular, and we obtain what is called an *equiangular* or *rectangular* hyperbola. A simple example is the *unit hyperbola*

$$x^2 - y^2 = 1. \tag{5.24}$$

Recalling that $x^2 - y^2 = (x + y)(x - y)$, we can make a transformation of coordinates in which $(x + y) \to x$ and $(x - y) \to y$. This simplifies the equation to

$$xy = 1. \tag{5.25}$$

The hyperbola has been rotated by $45°$ and the asymptotes have become the coordinate axes themselves. These rectangular hyperbolas are shown in Figure 5.7. More generally, hyperbolas of the form $xy = $ const represent relations in which y is inversely proportional to x. An important example is Boyle's law relating the pressure and volume of an ideal gas at constant temperature, which can be expressed as $pV = $ const.

We have already considered parabolas of the form $(y - y_0) = k(x - x_0)^2$, in connection with the quadratic formula. Analogous "sideways" parabolas can be obtained when the roles of x and y are reversed. Parabolas, as well as ellipses and hyperbolas, can be oriented obliquely to the axes by appropriate choices of B, the coefficient of xy, in the conic-section equation (5.19), the simplest example being the $45°$ hyperbola

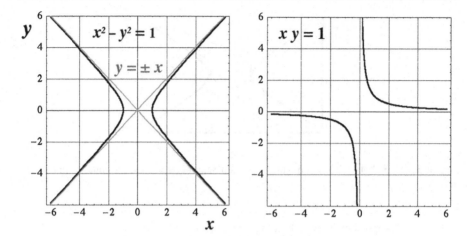

Figure 5.7 Rectangular hyperbolas.

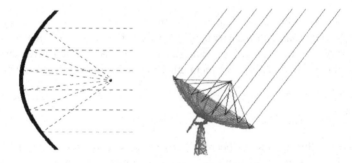

Figure 5.8 *Left:* Focusing of parallel lines by a parabola. *Right:* Paraboloid dish antenna.

$xy = 1$ considered above. Parabolas have the unique property that parallel rays incident upon them are reflected to a single point, called the *focus*, as shown in Figure 5.8. A parabola with the equation

$$y^2 = 4px \tag{5.26}$$

has its focus at the point $(p, 0)$. A parabola rotated about its symmetry axis generates a *paraboloid*. This geometry is exploited in applications where radiation needs to be concentrated at one point, such as radio telescopes, television dishes, and solar radiation collectors; also, when light emitted from a single point is to be projected as a parallel beam, as in automobile headlight reflectors.

Problem 5.3.1.   Find the center and radius of the circle with equation $Ax^2 + Ay^2 + Bx + Cy + D = 0$.

Problem 5.3.2.   Propose equations of three full circles and two semicircles which can be used to construct a Yin-Yang symbol.

Problem 5.3.3. The *latus rectum* (pardon the expression!) of a conic section is the length of a chord parallel to the directrix. Find the latus rectum of the ellipse $\frac{(x-x_0)^2}{a^2} + \frac{(y-y_0)^2}{b^2} = 1$ and of the parabola $y^2 = 4px$.

Problem 5.3.4. Find the four points of intersection of the confocal ellipse $\frac{x^2}{a^2} + \frac{y^2}{b^2} = 1$ and hyperbola $\frac{x^2}{a^2} - \frac{y^2}{b^2} = 1$.

5.4 Conic Sections in Polar Coordinates

The equations for conic sections can be expressed rather elegantly in polar coordinates. As shown in Figure 5.9, the origin is defined as the focus and a line corresponding to $x = d$ serves as the *directrix*. Recall the relations between the Cartesian and polar coordinates: $x = r\cos\theta$, $y = r\sin\theta$. The point P will trace out the conic section, moving in such a way that the ratio of its distance to the focus r to its distance to the

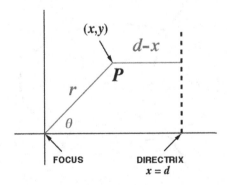

Figure 5.9 Coordinates used to represent conic sections. The point P traces out a conic section as r is varied, keeping a constant value of the eccentricity.

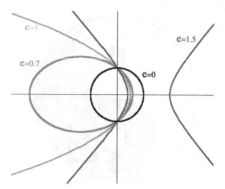

Figure 5.10 Polar plots of the conic sections, $r = 1/(1 + e\cos\theta)$, showing circle ($e = 0$), ellipse ($e = 0.7$), parabola ($e = 1$), and hyperbola ($e = 1.5$).

directrix $d - x = d - r\cos\theta$ is a constant. This ratio is called the *eccentricity*, e (not to be confused with Euler's $e = 2.718\ldots$):

$$e = \frac{r}{d - x} = \frac{r}{d - r\cos\theta}. \tag{5.27}$$

The polar equation for a conic section is then found by solving for r:

$$r = \frac{p}{1 + e\cos\theta}, \tag{5.28}$$

where the product ed is most conveniently replaced by a single constant p. The equation for a circle around the origin is simply $r = $ const. Thus a circle has $e = 0$ (think of $\bigcirc$). Eccentricity can, in fact, be thought of as a measure of how much a conic section deviates from being circular. For $0 < e < 1$, Eq. (5.28) represents an ellipse, for $e = 1$, a parabola and for $e > 1$, a hyperbola. Conic sections corresponding to

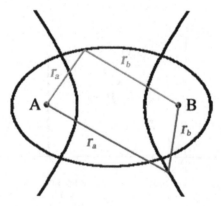

Figure 5.11 Confocal ellipse and hyperbola. Foci at points A, B. Ellipse is locus of constant $r_a + r_b$, hyperbola, locus of points of constant $|r_a - r_b|$.

several values of e are shown in Figure 5.10. As $e \to \infty$, the curve degenerates into a straight line. Newton showed that, under the inverse-square attraction of gravitational forces, the motion of a celestial object follows the trajectory of a conic section. The stable orbits of the planets around the Sun are ellipses, as found by Kepler's many years of observation of planetary motions. A parabolic or hyperbolic trajectory would represent a single pass through the Solar System, possibly that of a comet. The better-known comets have large elliptical orbits with eccentricities close to 1 and thus have long intervals between appearances. Halley's comet, for example, has $e = 0.967$ and a period of 76 years.

Ellipses and hyperbolas clearly have *two* distinct foci. The same ellipse or hyperbola can be constructed using its other focus and a corresponding directrix. In terms of their semimajor and semiminor axes, the eccentricities of ellipses and hyperbolas are given by

$$\text{ellipse}: \ e = \sqrt{1 - \frac{b^2}{a^2}}, \quad \text{hyperbola}: \ e = \sqrt{1 + \frac{b^2}{a^2}}. \tag{5.29}$$

Another way of constructing ellipses and hyperbolas makes use of their two foci, labeled A and B. An ellipse is the locus of points the sum of whose distances to the two foci, $r_a + r_b$, has a constant value. Different values of the sum generate a family of ellipses. Analogously, a hyperbola is the locus of points such that the *difference* $|r_a - r_b|$ is constant. The two families of *confocal* ellipses and hyperbolas are mutually orthogonal—that is, every intersection between an ellipse and a hyperbola meets at a 90° angle. This is shown in Figure 5.11.

Problem 5.4.1. Kepler's law of planetary motion for the orbit of a planet around the Sun can be expressed as

$$\frac{1}{r} = \frac{GMm^2}{L^2} \left(1 + \sqrt{1 + \frac{2EL^2}{G^2M^2m^3}} \cos\theta \right),$$

where G is the gravitational constant, M, the mass of the Sun, m, the mass of the planet, and E and L, the energy and angular momentum of the orbit. In terms of these variables, determine the eccentricity of the orbit, the perihelion (distances of closest approach) and the aphelion (farthest distance of the planet from the Sun).

6 Calculus

Knowledge of the calculus is often regarded as the dividing line between amateur and professional scientists. Calculus is regarded, in its own right, as one of the most beautiful creations of the human mind, comparable in its magnificence with the masterworks of Shakespeare, Mozart, Rembrandt, and Michelangelo. The invention of calculus is usually credited to Isaac Newton and Gottfried Wilhelm Leibniz in the 17th century. Some of the germinal ideas can, however, be traced back to Archimedes in the third century BC. Archimedes exploited the notion of adding up an infinite number of infinitesimal elements in order to determine areas and volumes of geometrical figures. We have already mentioned how he calculated the value of π by repeatedly doubling the number of sides of a polygon inscribed in a circle. The prototype problem in differential calculus is to determine the *slope* of a function $y(x)$ at each point x. As we have seen, this is easy for a straight line. The challenge comes when the function has a more complicated dependence on x. A further elaboration concerns the *curvature* of a function, describing how the slope changes with x. Newton's motivation for inventing differential calculus was to formulate the laws of motion—to determine how the planets move under the gravitational attraction of the Sun, how the Moon moves around the Earth, and how fast an apple falls to the ground from a tree. Thus Newton was most directly concerned with how quantities change as functions of *time*, thereby involving quantities such as velocity and acceleration.

6.1 A Little Road Trip

What does it mean when the speedometer on your car reads 35 miles/hr at some particular instant? It *doesn't* mean, you will readily agree, that you have come exactly 35 miles in the last hour or that you can expect to travel 35 miles in the next hour. You will almost certainly slow down and speed up during different parts of your trip and your speedometer will respond accordingly. You can certainly calculate your *average*

Guide to Essential Math 2e. http://dx.doi.org/10.1016/B978-0-12-407163-6.00006-0

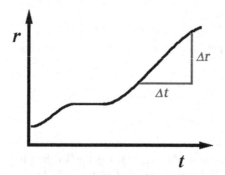

Figure 6.1 Plot of road trip. Speedometer reading gives $\lim_{\Delta t \to 0} \Delta r / \Delta t$.

speed for the entire journey by dividing the number of miles by the number of hours, but your speedometer readings will have been sometimes slower, sometimes faster, than this average value.

Let the variable r represent the distance you have traveled from your starting point, and let t represent the elapsed time. Figure 6.1 is a plot of your progress, distance traveled as a function of time, as represented by the function $r(t)$. Your stops for red lights show up as horizontal segments, where t continues to increase but r stands still. Suppose, at the beginning of your trip, your watch reads t_a and your odometer reads r_a, while, at the end, your watch and odometer read t_b and r_b, respectively. Your average speed—we'll call it v for velocity—for the whole trip is given by

$$v_{ab} = \frac{r_b - r_a}{t_b - t_a}. \tag{6.1}$$

You might have noted that the later part of your trip, after odometer reading r_c at time t_c, was somewhat faster than the earlier part. You could thereby calculate your speeds for the separate legs of the trip

$$v_{ac} = \frac{r_c - r_a}{t_c - t_a}, \quad v_{cb} = \frac{r_b - r_c}{t_b - t_c}. \tag{6.2}$$

You might continue dividing your trip into smaller and smaller increments, and calculate your average speed for each increment. Eventually, you should be able to match the actual readings on your speedometer!

To do this more systematically, let us calculate the average speed over a small time interval Δt around some time t, say between $t - \frac{1}{2}\Delta t$ and $t + \frac{1}{2}\Delta t$. We wind up with the same result somewhat more neatly by considering the time interval between t and $t + \Delta t$. Let the corresponding odometer readings be designated $r(t)$ and $r(t + \Delta t)$ and their difference by $\Delta r = r(t + \Delta t) - r(t)$. The average speed in this interval is given by

$$v(t) \approx \frac{\Delta r}{\Delta t} = \frac{r(t + \Delta t) - r(t)}{\Delta t} \tag{6.3}$$

and is represented by the slope of the chord intersecting the curve at the two points r, t and $r + \Delta r, t + \Delta t$. As we make Δt and Δr smaller and smaller, the secant will approach the tangent to the curve at the point r, t. The slope of this tangent then represents the instantaneous speed—as shown by the speedometer reading—at time t. This can be expressed mathematically as

$$v(t) = \lim_{\Delta t \to 0} \left[\frac{r(t + \Delta t) - r(t)}{\Delta t} \right]. \tag{6.4}$$

In the notation of differential calculus, this limit is written

$$v(t) = \frac{dr}{dt} \tag{6.5}$$

verbalized as "the derivative of r with respect to t" or more briefly as "DRDT." Alternative ways of writing the derivative are dr/dt, $r'(t)$, and $\frac{d}{dt} r(t)$. For the special case when the independent variable is time, its derivative, the velocity, is written $\dot{r}(t)$. This was Newton's original notation for the quantity he called a "fluxion."

You've possibly heard about a hot new Porsche that can "accelerate from 0 to 60 mph in 3.8 seconds." Just as velocity is the time derivative of distance, acceleration is the time derivative of velocity:

$$a(t) = \frac{dv}{dt}. \tag{6.6}$$

It thus represents the *second derivative* of distance, written

$$a(t) = \frac{d}{dt}\left(\frac{dr}{dt}\right) \equiv \frac{d^2 r}{dt^2}. \tag{6.7}$$

Alternative notations for $a(t)$ are $r''(t)$ and $\ddot{r}(t)$. Newton's second law of motion states that the force F on a body of mass m causes an acceleration given by

$$F = ma. \tag{6.8}$$

6.2   A Speedboat Ride

After your drive to your lakeside destination, you might want to take a spin in your new speedboat. Speedboats are likely to have speedometers but not odometers. Suppose, given your newfound appreciation of calculus, you would like to somehow apply calculus to your speedboat ride. It turns out that using data from your speedometer and wristwatch, you can determine the *distance* your boat has traveled. Dimensionally, distance = speed × time, or, expressed in the style of factor-label analysis,

$$\text{miles} = \frac{\text{miles}}{\boxed{\text{h}}} \times \boxed{\text{h}}. \tag{6.9}$$

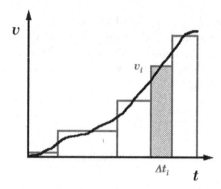

Figure 6.2 Plot of velocity vs. time. Distance traveled is approximated by summing over rectangular strips.

During a short interval of time Δt_i around a time t_i, your velocity might be practically constant, say, $v(t_i)$ mph. The distance you covered during this time would then be given by $v(t_i)\Delta t_i$. If you start at time t_a, the distance you cover by time t_b, namely $r(t_b) - r(t_a)$, can be approximated by the sum of n individual contributions:

$$r(t_b) - r(t_a) \approx \sum_{i=1}^{n} v(t_i)\Delta t_i. \tag{6.10}$$

This can be represented, as shown in Figure 6.2, as the sum of *areas* of a series of vertical strips of height $v(t_i)$ and width Δt_i. In concept, your computation of distance can be made exact by making the time intervals shorter and shorter ($\Delta t_i \to 0$ for all i) and letting the number of intervals approach infinity ($n \to \infty$). Graphically, this is equivalent to finding the area under a smooth curve representing $v(t)$ between the times t_a and t_b. This defines the *definite integral* of the function $v(t)$, written

$$r(t_b) - r(t_a) = \int_{t_a}^{t_b} v(t)dt \equiv \lim_{\substack{n \to \infty \\ \Delta t_i \to 0}} \sum_{i=1}^{n} v(t_i)\Delta t_i. \tag{6.11}$$

6.3 Differential and Integral Calculus

Let us reiterate the results of the last two sections using more standard notation. Expressed in the starkest terms, the two fundamental operations of calculus have the objective of either (i) determining the slope of a function at a given point or (ii) determining the area under a curve. The first is the subject of *differential calculus*, the second, of *integral calculus*.

Consider a function $y = F(x)$, which is graphed in Figure 6.3. The slope of the function at the point x can be determined by a limiting process in which a small chord through the points $x, F(x)$ and $x + \Delta x, F(x + \Delta x)$ is made to approach the tangent at $x, F(x)$. The slope of this tangent is understood to represent the slope of the function $F(x)$ at the point x. Its value is given by the derivative

$$\frac{dy}{dx} = \lim_{\Delta x \to 0} \left[\frac{F(x + \Delta x) - F(x)}{\Delta x} \right], \tag{6.12}$$

which can also be written dF/dx, $F'(x)$, or $y'(x)$. When Δx, a small increment of x, approaches zero, it is conventionally written dx, called the *differential* of x. Symbolically:

$$\text{As} \quad \Delta x \to 0, \quad \Delta x \Rightarrow dx. \tag{6.13}$$

Note that the limit in Eq. (6.12) involves the ratio of two quantities *both* of which approach zero. It is an article of faith to accept that their ratio can still approach a finite limit while both numerator and denominator vanish. In the words of Bishop Berkeley, a contemporary of Newton, "May we not call them ghosts of departed quantities?"

The prototype problem in integral calculus is to determine the area under a curve representing a function $f(x)$ between the two values $x = a$ and $x = b$, as shown in Figure 6.4. The strategy again is to approximate the area by an array of rectangular strips. It is most convenient to divide the range $a \leqslant x \leqslant b$ into n strips of equal width Δx. We use the convention that the ith strip lies between the values labeled x_i and x_{i+1}. Consistent with this notation, $x_0 = a$ and $x_n = b$. Also note that $x_{i+1} - x_i = \Delta x$ for

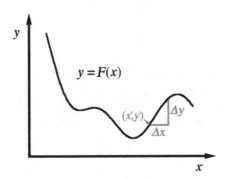

Figure 6.3 Graph of the function $y = F(x)$. The ratio $\Delta y/\Delta x$ approximates the slope at the point (x', y').

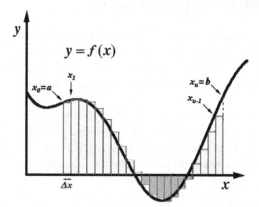

Figure 6.4 Evaluation of the definite integral $\int_a^b f(x)dx$. The areas of the rectangles above the x-axis are added, those below the x-axis are subtracted. The integral equals the limit as $n \to \infty$ and $\Delta x \to 0$.

all i. The area of the n strips adds up to

$$\sum_{i=0}^{n-1} f(x_i)\Delta x. \tag{6.14}$$

In mathematical jargon, this is called a *Riemann sum*. As we divide the area into a greater and greater number of narrower strips, $n \to \infty$ and $\Delta x \to 0$. The limiting process defines the *definite integral* (also called a *Riemann integral*):

$$\int_a^b f(x)dx \equiv \lim_{\substack{n\to\infty \\ \Delta x \to 0}} \sum_{i=1}^{n-1} f(x_i)\Delta x. \tag{6.15}$$

Note that when the function $f(x)$ is negative, it *subtracts* from the sum (6.14). Thus the integral (6.15) represents the *net* area above the x-axis, with regions below the axis making negative contributions.

Suppose now that the function $f(x)$ has the property that $F'(x) = f(x)$, where the function $F(x)$ is called the *antiderivative* of $f(x)$. Accordingly, $f(x_i)$ in Eq. (6.14) can be approximated by

$$f(x_i) = F'(x_i) \approx \frac{F(x_i + \Delta x) - F(x_i)}{\Delta x}. \tag{6.16}$$

Noting that $x_i + \Delta x = x_{i+1}$, Eq. (6.14) can be written

$$\sum_{i=0}^{n-1} f(x_i)\,\Delta x = \sum_{i=0}^{n-1} \frac{F(x_i + \Delta x) - F(x_i)}{\Delta x}\,\Delta x$$

$$= [\,F(x_1) - F(x_0)\,] + [\,F(x_2) - F(x_1)\,]$$

$$+ [\,F(x_3) - F(x_2)\,] + \cdots + [F(x_n) - F(x_{n-1})\,]$$

$$= F(x_n) - F(x_0) = F(b) - F(a) \qquad (6.17)$$

Note that every intermediate value $F(x_i)$ is canceled out in successive terms.

In the limit as $n \to \infty$ and $\Delta x \to 0$, we arrive at the *fundamental theorem of calculus*:

$$\int_a^b f(x)\,dx = F(x)\Big|_a^b = F(b) - F(a), \qquad \text{where } F'(x) = f(x). \qquad (6.18)$$

This connects differentiation with integration and shows them to be essentially inverse operations.

In our definitions of derivatives and integrals, we have been carefree in assuming that the functions $F(x)$ and $f(x)$ were appropriately well behaved. For functions which correspond to physical variables, this is almost always the case. But just to placate any horrified mathematicians who might be reading this, there are certain conditions which must be fulfilled for functions to be differentiable and/or integrable. A necessary condition for $F'(x)$ to exist is that the function be *continuous*. Figure 6.5 shows an example of a function $F(x)$ with a discontinuity at $x = a$. The derivative cannot be defined at that point. (Actually, for a finite-jump discontinuity, mathematical physicists regard $F'(x)$ as proportional to the *deltafunction*, $\delta(x-a)$, which has the remarkable property of being infinite at the point $x = a$, but zero everywhere else.) Even a continuous function can be nondifferentiable, for example, the function $\sin(1/x)$, which oscillates so rapidly as $x \to 0$ that its derivative at $x = 0$, is undefined. Such pathological behavior is, as we have noted, rare in physical applications. We might also have to contend with functions which are continuous but not *smooth*. In such cases, the derivative $F'(x)$ at a point is discontinuous, depending on which direction it is evaluated. The prototype example is the absolute value function $F(x) = |x|$. For $x < 0$, $F'(x) = -1$, while for $x > 0$, $F'(x) = +1$, thus the derivative is discontinuous at $x = 0$.

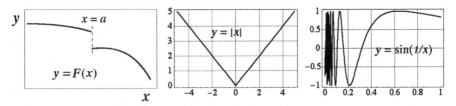

Figure 6.5 Three functions with pathologies in their derivatives. *Left:* $y = F(x)$ is discontinuous at $x = a$. *Center:* $y = |x|$ has discontinuous derivative at $x = 0$. *Right:* $y = \sin(1/x)$ has undefined derivative as $x \to 0$.

Generally, the definite integral exists for functions that have at most a finite number of finite discontinuities—classified as *piecewise continuous*. Most often an integral "does not exist" if it blows up to an infinite value, for example, $\int_0^1 x^{-1}\,dx$. These are also known as *improper integrals*.

6.4 Basic Formulas of Differential Calculus

The terms "differentiating" and "finding the derivative" are synonymous. A few simple rules suffice to determine the derivatives of most functions you will encounter. These can usually be deduced from the definition of derivative in Eq. (6.12). Consider first the function $y(x) = ax^n$, where a is a constant. We will need

$$y(x + \Delta x) = a(x + \Delta x)^n = a\left[x^n + nx^{n-1}\Delta x + \frac{n(n-1)}{2}x^{n-2}(\Delta x)^2 + \cdots\right]$$

(6.19)

from the binomial expansion (3.92). It follows then that

$$\frac{y(x + \Delta x) - y(x)}{\Delta x} = a\left[nx^{n-1} + \frac{n(n-1)}{2}x^{n-2}\Delta x + \cdots\right].$$

(6.20)

Finally, taking the limit $\Delta x \to 0$, we find

$$\frac{d}{dx}(ax^n) = anx^{n-1}.$$

(6.21)

For the cases $n = 0, 1, 2$:

$$\frac{d}{dx}a = 0, \quad \frac{d}{dx}(ax) = a, \quad \frac{d}{dx}(ax^2) = 2ax.$$

(6.22)

The first formula means that the derivative of a constant is zero. Equation (6.21) is also valid for fractional or negative values of n. Thus we find

$$\frac{d}{dx}\left(\frac{1}{x}\right) = -\frac{1}{x^2}, \quad \frac{d}{dx}\sqrt{x} = \frac{1}{2\sqrt{x}}, \quad \frac{d}{dx}x^{-1/2} = -\frac{x^{-3/2}}{2}.$$

(6.23)

For the exponential function $y(x) = e^x$, we find

$$\frac{y(x + \Delta x) - y(x)}{\Delta x} = \frac{e^x e^{\Delta x} - e^x}{\Delta x}$$

$$= \frac{e^x}{\Delta x}\left(1 + \Delta x + \frac{(\Delta x)^2}{2} + \cdots\right) - \frac{e^x}{\Delta x}$$

$$= e^x\left(1 + \frac{\Delta x}{2} + \cdots\right).$$

(6.24)

In the limit $\Delta x \to 0$, we find

$$\frac{d}{dx}e^x = e^x. \tag{6.25}$$

Thus the exponential function equals its own derivative! This result also follows from term-by-term differentiation of the series (3.104). The result (6.25) is easy generalized to give

$$\frac{d}{dx}(ae^{cx}) = ace^{cx}. \tag{6.26}$$

For the natural logarithm $y(x) = \ln x$, we find

$$\frac{y(x + \Delta x) - y(x)}{\Delta x} = \frac{\ln(x + \Delta x) - \ln x}{\Delta x} = \frac{1}{\Delta x}\ln\left[\left(\frac{x + \Delta x}{x}\right)\right]$$

$$= \ln\left[\left(1 + \frac{\Delta x}{x}\right)^{1/\Delta x}\right] \xrightarrow{\Delta x \to 0} \ln(e^{1/x}) = \frac{1}{x}. \tag{6.27}$$

having used several properties of logarithms and the definition of the exponential function. Therefore

$$\frac{d}{dx}\ln x = \frac{1}{x}. \tag{6.28}$$

For logarithm to the base b

$$\frac{d}{dx}\log_b x = \frac{1}{x \ln b}. \tag{6.29}$$

We can thus show

$$\frac{d}{dx}b^x = \frac{d}{dx}e^{x \ln b} = \ln b\, e^{x \ln b} = b^x \ln b. \tag{6.30}$$

Don't confuse this with the result $dx^b/dx = b x^{b-1}$.

Derivatives of the trigonometric functions can be readily found using Euler's theorem (4.57):

$$\frac{d}{dx}e^{ix} = \frac{d}{dx}\cos x + i\frac{d}{dx}\sin x. \tag{6.31}$$

Therefore

$$ie^{ix} = i(\cos x + i \sin x) = \frac{d}{dx}\cos x + i\frac{d}{dx}\sin x \tag{6.32}$$

and equating the separate real and imaginary parts, we find

$$\frac{d}{dx}\sin x = \cos x \quad \text{and} \quad \frac{d}{dx}\cos x = -\sin x. \tag{6.33}$$

The other trigonometric derivatives can be found from these and we simply list the results:

$$\frac{d}{dx}\tan x = \sec^2 x, \quad \frac{d}{dx}\cot x = -\csc^2 x,$$
$$\frac{d}{dx}\sec x = \sec x \tan x, \quad \frac{d}{dx}\csc x = -\csc x \cot x. \tag{6.34}$$

The derivatives of the hyperbolic functions are easily found from their exponential forms (4.66). These are analogous to the trigonometric results, except that there is no minus sign:

$$\frac{d}{dx}\sinh x = \cosh x \quad \text{and} \quad \frac{d}{dx}\cosh x = \sinh x. \tag{6.35}$$

6.5   More on Derivatives

Techniques which enable us to find derivatives of more complicated functions can be based on the *chain rule*. Suppose we are given what can be called a "function of a function of x," say $g[f(x)]$. For example, the Gaussian function e^{-x^2} represents an exponential of the square of x. The derivative of a composite function involves the limit of the quantity

$$\frac{g[f(x+\Delta x)] - g[f(x)]}{\Delta x}.$$

The function $f(x)$ can be considered a variable itself, in the sense that $\Delta f = f(x + \Delta x) - f(x)$. We can therefore write

$$\frac{g[f(x+\Delta x)] - g[f(x)]}{\Delta x} = \frac{g(f - \Delta f)}{\Delta f}\frac{\Delta f}{\Delta x} \xrightarrow{\Delta x \to 0} g'[f(x)]f'(x). \tag{6.36}$$

For example,

$$\frac{d}{dx}e^{-x^2} = e^{-x^2}\frac{d}{dx}(-x^2) = -2xe^{-x^2}. \tag{6.37}$$

In effect, we have evaluated this derivative by a change of variables from x to $-x^2$.

The derivatives of the inverse trigonometric functions, such as $d\arcsin x/dx$, can be evaluated using the chain rule. If $\theta(x) = \arcsin x$, then $x = \sin[\theta(x)]$. Taking d/dx of both sides in the last form, we find

$$\frac{d}{dx}x = \frac{d}{dx}\sin[\theta(x)] = \frac{d\sin\theta}{d\theta}\frac{d\theta(x)}{dx} = \cos\theta\frac{d\theta}{dx}. \tag{6.38}$$

But

$$\cos\theta = \sqrt{1 - \sin^2\theta} = \sqrt{1 - x^2}, \tag{6.39}$$

so that

$$\frac{d}{dx}\arcsin x = \frac{1}{\sqrt{1-x^2}}.$$

(6.40)

We can show analogously that

$$\frac{d}{dx}\arccos x = -\frac{1}{\sqrt{1-x^2}}$$

(6.41)

and

$$\frac{d}{dx}\arctan x = \frac{1}{1+x^2}.$$

(6.42)

Implicit differentiation is a method of finding dy/dx for a functional relation which cannot be easily solved for $y(x)$. Suppose, for example, we have y and x related by $y = xe^y$. This cannot be solved for y in closed form. However, taking d/dx of both sides and solving for dy/dx, we obtain

$$\frac{dy}{dx} = e^y + xe^y\frac{dy}{dx}, \quad \frac{dy}{dx} = \frac{e^y}{1-xe^y} = \frac{e^y}{1-y}.$$

(6.43)

Implicit differentiation can be applied more generally to any functional relation of the form $f(x, y) = 0$.

We have already used the fact that the derivative of a sum or difference equals the sum or difference of the derivatives. More generally

$$\frac{d}{dx}[af(x) + bg(x) + \cdots] = a\frac{d}{dx}f(x) + b\frac{d}{dx}g(x) + \cdots$$

(6.44)

The derivative of a product of two functions is given by

$$\frac{d}{dx}[f(x)g(x)] = f(x)\frac{d}{dx}g(x) + g(x)\frac{d}{dx}f(x) = f(x)g'(x) + g(x)f'(x),$$

(6.45)

while for a quotient

$$\frac{d}{dx}\frac{f(x)}{g(x)} = \frac{g(x)f'(x) - f(x)g'(x)}{g(x)^2}.$$

(6.46)

Problem 6.5.1. Hermite polynomials can be defined in terms of multiple derivatives as follows:

$$H_n(x) = (-1)^n e^{x^2}\frac{d^n}{dx^n}e^{-x^2}.$$

Calculate $H_n(x)$ for $n = 0$, 1, and 2 ($n = 0$ means no differentiation).

Problem 6.5.2. Analogously, Legendre polynomials can be defined by

$$P_n(x) = \frac{1}{2^n n!}\frac{d^n}{dx^n}(1 - x^2)^n.$$

Calculate $P_n(x)$ for $n = 0$, 1, and 2.

6.6 Indefinite Integrals

We had earlier introduced the *antiderivative* $F(x)$ of a function $f(x)$, meaning that $F'(x) = f(x)$. Since $F(x) = x^n$ gives $f(x) = nx^{n-1}$, the inverse would imply that $F'(x) = x^m$ must be the derivative of $f(x) = x^{m+1}/(m+1)$. More generally, we could say that the antiderivative of x^m equals $x^{m+1}/(m+1)+$const since the constant will disappear upon taking the derivative.

The fundamental theorem (6.18) can be rewritten with x' replacing x as the integration variable and x replacing the limiting value b. This gives

$$\int_a^x f(x')dx' = F(x) - F(a). \tag{6.47}$$

This will be expressed in the form

$$F(x) = \int f(x)dx + \text{const.} \tag{6.48}$$

The antiderivative of a function $f(x)$ will hereafter be called the *indefinite integral* and be designated $\int f(x)dx$. Thus the result derived in the last paragraph can now be written

$$\int x^m \, dx = \frac{x^{m-1}}{m-1} + \text{const.} \tag{6.49}$$

All the derivatives we obtained in Sections 6.4 and 6.5 can now be "turned inside out" to give the following integral formulas; in all cases a constant is to be added to the right-hand side:

$$\int e^x dx = e^x, \tag{6.50}$$

$$\int e^{cx} dx = \frac{e^{cx}}{c}, \tag{6.51}$$

$$\int \frac{1}{x}dx = \int \frac{dx}{x} = \ln x, \tag{6.52}$$

$$\int \sin x dx = -\cos x, \tag{6.53}$$

$$\int \cos x dx = \sin x, \tag{6.54}$$

$$\int \frac{dx}{1+x^2} = \arctan x, \tag{6.55}$$

$$\int \frac{dx}{\sqrt{1-x^2}} = \arcsin x. \tag{6.56}$$

For all the above integrals, the constant drops out if we put in limits of integration, for example

$$\int_a^b \frac{dx}{x} = \ln x \Big|_a^b = \ln\left(\frac{b}{a}\right). \tag{6.57}$$

You can find many Tables of Integrals which list hundreds of other functions. A very valuable resource is the *Mathematica* integration website: http://integrals. wolfram.com/. For example, you can easily find that

$$\int \frac{x}{\sqrt{a^2 - x^2}}\, dx = -\sqrt{a^2 - x^2} + \text{const.} \tag{6.58}$$

You do have to use the *Mathematica* conventions for the integrand, in this case "x/Sqrt[a^2 - x^2]."

From a fundamental point of view, integration is less demanding than differentiation, as far as the conditions imposed on the class of functions. As a consequence, numerical integration is a lot easier to carry out than numerical differentiation. If we seek explicit functional forms (sometimes referred to as *closed forms*) for the two operations of calculus, the situation is reversed. You can find a closed form for the derivative of almost any function. But, even some simple functional forms cannot be integrated explicitly, at least not in terms of elementary functions. For example, there are no simple formulas for the indefinite integrals $\int e^{-x^2}\, dx$ or $\int \frac{e^{-x}}{x}\, dx$. These can, however, be used to definite new functions, namely, the error function and the exponential integral for the two examples just given.

6.7 Techniques of Integration

There are a number of standard procedures which can enable a large number of common integrals to be evaluated explicitly. The simplest strategy is *integration by substitution*, which means changing of the variable of integration. Consider, for example, the integral $\int xe^{-x^2}\, dx$. The integral can be evaluated in closed form even though $\int e^{-x^2}\, dx$ cannot. The trick is to define a new variable $y = x^2$, so that $x = \sqrt{y}$. We have then that $dx = dy/2\sqrt{y}$. The integral becomes tractable in terms of y:

$$\int xe^{-x^2}\, dx = \int \sqrt{y}e^{-y}\frac{dy}{2\sqrt{y}} = \frac{1}{2}\int e^{-y}\, dy = -\frac{1}{2}e^{-y} = -\frac{1}{2}e^{-x^2}. \tag{6.59}$$

The result can be checked by taking the derivative of e^{-x^2}.

As a second example consider the integral (6.58) above, which we found using the *Mathematica* computer program. A first simplification would be to write $x = ay$ so that

$$\int \frac{x}{\sqrt{a^2 - x^2}}\, dx = a \int \frac{y}{\sqrt{1 - y^2}}\, dy. \tag{6.60}$$

Next we note the tantalizing resemblance of $\sqrt{1 - y^2}$ to $\sqrt{1 - \sin^2 \theta}$. This suggests a second variable transformation $y = \sin \theta$, with $dy = \cos \theta \, d\theta$. The integral becomes

$$a \int \frac{\sin \theta \cos \theta}{\sqrt{1 - \sin^2 \theta}}\, d\theta = a \int \sin \theta \, d\theta = -a \cos \theta = -a\sqrt{1 - \sin^2 \theta}$$

$$= -a\sqrt{1 - y^2} = -\sqrt{a^2 - x^2} \tag{6.61}$$

in agreement with the result obtained earlier.

Trigonometric identities suggest that integrals containing the forms $\sqrt{a^2 - x^2}$ can sometimes be evaluated using a substitution $x = a \sin \theta$ or $x = a \cos \theta$. Likewise forms containing $\sqrt{x^2 - a^2}$ suggest a possible transformation such as $x = a \csc \theta$ or $x = a \cosh t$. Forms containing $\sqrt{x^2 + a^2}$ suggest possibilities such as $x = a \tan \theta$ or $x = a \sinh t$.

Integration by parts is another method suggested by the formula for the derivative of a product, Eq. (6.45). In differential form, this can be expressed

$$d(uv) = u \, dv + v \, du, \tag{6.62}$$

where u and v are understood to be functions of x. Integrating (6.62). we obtain the well-known formula for integration by parts

$$\int u \, dv = uv - \int v \, du. \tag{6.63}$$

This is useful whenever $\int v \, du$ is easier to evaluate than $\int u \, dv$. As an example, consider $\int \ln x \, dx$, another case of a very elementary function which doesn't have an easy integral. But if we set $u = \ln x$ and $v = x$, then $du = dx/x$ and we find using (6.63) that

$$\int \ln x \, dx = x \ln x - \int x \frac{dx}{x} = x \ln x - x. \tag{6.64}$$

An integral of the type

$$\int \frac{f(x)}{(x-a)(x-b)} \, dx$$

can often be evaluated by the *method of partial fractions*. We can always find constants A and B such that

$$\frac{1}{(x-a)(x-b)} = \frac{A}{x-a} + \frac{B}{x-b}. \tag{6.65}$$

Therefore the integral can be reduced to

$$\int \frac{f(x)}{(x-a)(x-b)} \, dx = \int \frac{f(x)}{x-a} \, dx + \int \frac{f(x)}{x-b} \, dx, \tag{6.66}$$

which might be easier to solve. In the event that the denominator contains factors raised to powers, the procedure must be generalized. For example,

$$\frac{1}{(x-a)^2(x-b)} = \frac{A}{(x-a)^2} + \frac{B}{x-a} + \frac{C}{x-b} \tag{6.67}$$

and more generally,

$$\frac{1}{(x-a)^N(x-b)^M} = \sum_{n=1}^{N} \frac{A_n}{(x-a)^n} + \sum_{m=1}^{M} \frac{B_m}{(x-b)^m}. \tag{6.68}$$

Problem 6.7.1. Evaluate the integral

$$\int \frac{x}{\sqrt{a^2 + x^2}} \, dx.$$

Problem 6.7.2. Evaluate the integral $\int x \sin x \, dx$ using integration by parts.

Problem 6.7.3. Evaluate the integral $\int x^2 \sin x \, dx$. This will involve integration by parts twice.

Problem 6.7.4. Evaluate the integral

$$\int \frac{dx}{(x-a)^2(x-b)}$$

using the method of partial fractions.

6.8 Curvature, Maxima and Minima

The *second derivative* of a function $f(x)$ is the derivative of $f'(x)$, defined by

$$f''(x) = \frac{d}{dx} f'(x) = \frac{d^2}{dx^2} f(x) = \lim_{\Delta x \to 0} \left[\frac{f'(x + \Delta x) - f'(x)}{\Delta x} \right]. \tag{6.69}$$

Putting in the definition (6.12) of the first derivative, this can also be written

$$f''(x) = \lim_{\Delta x \to 0} \left[\frac{f(x + 2\Delta x) - 2f(x + \Delta x) + f(x)}{(\Delta x)^2} \right]. \tag{6.70}$$

This formula is convenient for numerical evaluation of second derivatives. For analytical purposes, we can simply apply all the derivative techniques of Sections 6.4 and 6.5 to the function $f'(x)$. Higher derivatives can be defined analogously

$$f^{(n)}(x) \equiv \frac{d^n}{dx^n} f(x). \tag{6.71}$$

These will be used in the following chapter to obtain power-series representations for functions.

Recall that the first derivative $f'(x)$ is a measure of the instantaneous slope of the function $f(x)$ at x. When $f'(x) > 0$, the function is increasing with x, that is, it slopes *upward*. Conversely, when $f'(x) < 0$, the function decreases with x and slopes *downward*. At points x where $f'(x) = 0$ the function is instantaneously horizontal. This is called a *stationary point* and may represent a local maximum or minimum, depending on the sign of $f''(x)$ at that point.

The second derivative $f''(x)$ is analogously a measure of the increase or decrease in the slope $f'(x)$. When $f''(x) > 0$, the slope is increasing with x and the function has an upward *curvature*. It is concave upward and would hold water if it were a cup.

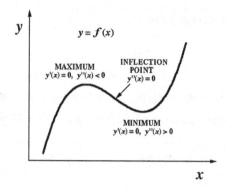

Figure 6.6 Maximum, minimum, and inflection points of function $y = f(x)$.

Conversely, when $f''(x) < 0$, the function must have a *downward* curvature. It is concave downward and water would spill out. A point where $f''(x) = 0$, where the curvature is zero, is known as an *inflection point*. Most often, for a continuous function, an inflection point represents a point of transition between positive and negative curvature.

Let us return to our consideration of stationary points, where $f'(x) = 0$. If $f''(x) < 0$, the curvature is downward and this must therefore represent a *local maximum* of the function $f(x)$. The tangent at the maximum rests *on top* of the curve. We call this maximum "local" because there is no restriction on $f(x)$ having an even larger value somewhere else. The maximum possible value of a function in its entire domain is called its *global maximum*. Analogously, when $f'(x) = 0$ and $f''(x) > 0$, we have a local minimum. In this case, the curve rests on its tangent. Three features described above are illustrated in Figure 6.6. A point where both $f'(x) = 0$ and $f''(x) = 0$, assuming the function is not simply a constant, is known as a *horizontal inflection point*.

6.9 The Gamma Function

The gamma function is one of a class of functions which is most conveniently defined by a definite integral. Consider first the following integral, which can be evaluated exactly:

$$\int_0^\infty e^{-\alpha x}\, dx = \frac{e^{-\alpha x}}{-\alpha} \Big|_0^\infty = \frac{1}{\alpha}. \tag{6.72}$$

A very useful trick is to take the derivative of an integral with respect to one of its parameters (not the variable of integration). Suppose we know the definite integral

$$\int_a^b f(\alpha, x)\, dx = F(\alpha, b) - F(\alpha, a), \tag{6.73}$$

where α is a parameter not involved in the integration. We can take $d/d\alpha$ of both sides to give

$$\frac{d}{d\alpha} \int_a^b f(\alpha, x)\, dx = \int_a^b \frac{\partial}{\partial \alpha} f(\alpha, x)\, dx = \frac{\partial}{\partial \alpha}[F(\alpha, b) - F(\alpha, a)]. \qquad (6.74)$$

This operation is valid for all reasonably well-behaved functions. (For the derivative of a function of two variables with respect to one of these variables, we have written the *partial derivative* $\partial/\partial\alpha$ in place of $d/d\alpha$. Partial derivatives will be dealt with more systematically in Chapter 10.) Applying this operation to the integral (6.72), we find

$$\frac{d}{d\alpha} \int_0^\infty e^{-\alpha x}\, dx = \int_0^\infty (-x)e^{-\alpha x}\, dx = \frac{d}{d\alpha}\left(\frac{1}{\alpha}\right) = -\frac{1}{\alpha^2}. \qquad (6.75)$$

We have therefore obtained a new definite integral:

$$\int_0^\infty x e^{-\alpha x}\, dx = \frac{1}{\alpha^2}. \qquad (6.76)$$

Taking $d/d\alpha$ again we find

$$\int_0^\infty x^2 e^{-\alpha x}\, dx = \frac{2}{\alpha^3}. \qquad (6.77)$$

Repeating the process n times

$$\int_0^\infty x^n e^{-\alpha x}\, dx = \frac{2 \cdot 3 \cdot 4 \cdots n}{\alpha^{n+1}}. \qquad (6.78)$$

Setting $\alpha = 1$, now that its job is done, we wind up a neat integral formula for $n!$

$$\int_0^\infty x^n e^{-x}\, dx = n! \qquad (6.79)$$

This is certainly not the most convenient way to evaluate $n!$, but suppose we replace n by a *noninteger* ν. In conventional notation, this defines the *gamma function*:

$$\Gamma(\nu) \equiv \int_0^\infty x^{\nu-1} e^{-x}\, dx. \qquad (6.80)$$

When ν is an integer, this reduces to the factorial by the relation

$$\Gamma(n+1) = n!, \quad n = 0, 1, 2, 3 \ldots \qquad (6.81)$$

Occasionally the notation $\nu!$ is used for $\Gamma(\nu + 1)$ even for noninteger ν.
   For the case $\nu = 1/2$

$$\Gamma\left(\frac{1}{2}\right) = \int_0^\infty x^{-1/2} e^{-x}\, dx. \qquad (6.82)$$

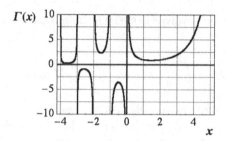

Figure 6.7 Plot of the gamma function.

The integral can be evaluated with a change of variables $x = y^2, dx = 2y\,dy$ giving

$$\Gamma\left(\frac{1}{2}\right) = 2\int_0^\infty e^{-y^2}\,dy = \sqrt{\pi},\tag{6.83}$$

where we have recalled Laplace's famous result from Eq. (1.21)

$$\int_{-\infty}^{\infty} e^{-x^2}\,dx = \sqrt{\pi} \Rightarrow \int_0^\infty e^{-x^2}\,dx = \frac{\sqrt{\pi}}{2}.\tag{6.84}$$

Thus

$$\left(-\frac{1}{2}\right)! = \Gamma\left(\frac{1}{2}\right) = \sqrt{\pi}\tag{6.85}$$

the relation we had teased you with in Eq. (3.84).

Figure 6.7 shows a plot of the gamma function. For $x > 0$, the function is a smooth interpolation between integer factorials. $\Gamma(x)$ becomes infinite for $x = 0, -1, -2, \ldots$

6.10   Gaussian and Error Functions

An apocryphal story is told of a math major showing a psychology student the formula for the infamous Gaussian or bell-shaped curve, which purports to represent the distribution of human intelligence and such. The formula for a normalized Gaussian looks like this:

$$P(x) = \frac{1}{\sigma\sqrt{2\pi}}e^{-(x-x_0)^2/2\sigma^2}\tag{6.86}$$

and is graphed in Figure 6.8. The psychology student, unable to fathom the fact that this formula contained π, the ratio between the circumference and diameter of a circle, asked, "Whatever does π have to do with intelligence?" The math student is supposed to have replied, "If your IQ were high enough, you would understand!" The π comes, of course, from Laplace's integral (1.21), slightly generalized to

$$\int_{-\infty}^{\infty} e^{-\alpha x^2}\,dx = \sqrt{\frac{\pi}{\alpha}}.\tag{6.87}$$

With the appropriate choice of variables, this gives the normalization condition for the Gaussian function

$$\int_{-\infty}^{\infty} P(x)\,dx = 1. \tag{6.88}$$

The average value of the variable x is given by

$$\overline{x} = \int_{-\infty}^{\infty} x P(x)\,dx = x_0. \tag{6.89}$$

The *standard deviation*, σ, commonly called "sigma," parametrizes the half-width of the distribution. It is defined as the *root mean square* of the distribution. The mean square is given by

$$\overline{(x - \overline{x})^2} = \int_{-\infty}^{\infty} (x - x_0)^2 P(x)\,dx = \sigma^2. \tag{6.90}$$

To evaluate the integrals (6.89) and (6.90) for the Gaussian distribution, we need the additional integrals

$$\int_{-\infty}^{\infty} x e^{-\alpha x^2}\,dx = 0 \quad\text{and}\quad \int_{-\infty}^{\infty} x^2 e^{-\alpha x^2}\,dx = \frac{\sqrt{\pi}}{2\alpha^{3/2}}. \tag{6.91}$$

Since the integrand in the first integral is an odd function, contributions from $x < 0$ and $x > 0$ exactly cancel to give zero. The second integral can be found by taking $d/d\alpha$ on both sides of (6.87), the same trick we used in Section 6.9. For the IQ distribution shown in Figure 6.8, the average IQ is 100 and sigma is approximately equal to 15 or 16 IQ points.

A Gaussian distribution can also arise as a limiting case of a binomial distribution. A good illustration is the statistics of coin tossing. Suppose that the toss of a coin gives, with equal *a priori* probability, heads (H) or tails (T). A second toss will give four equally possible results: HH, HT, TH, and TT, with a 1 2 1 distribution for 0, 1, and 2 heads, respectively. Three tosses will give eight equal possibilities: HHH, HHT, HTH, THH, TTH, THT, HTT, and TTT, with a 1 3 3 1 distribution for 0, 1, 2, and 3 heads, respectively. Clearly we are generating a binomial distribution of the form (3.91):

$$\binom{n}{r} = \frac{n!}{(n-r)!\,r!}, \tag{6.92}$$

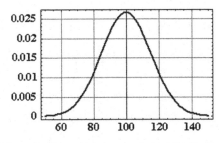

Figure 6.8 Normalized Gaussian applied to distribution of IQ's.

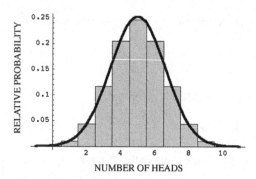

Figure 6.9 Binomial distribution for 10 coin tosses, shown as histogram. The distribution is well approximated by a Gaussian centered at $n = 5$ with $\sigma = \sqrt{5}/2$.

where r is the number of heads in n coin tosses. Figure 6.9 plots the binomial distribution for $n = 10$. As $n \to \infty$, the binomial distribution approaches a Gaussian (6.86), with $x_0 = n/2$ and $\sigma = \sqrt{n}/2$. Remarkably, sigma also increases with n but only as its square root. If we were to toss a coin one million times, the average number of heads would be 500,000 but the likely discrepancy would be around 500, one way or the other.

The percentage of a distribution between two finite values is obtained by integrating the Gaussian over this range:

$$P(a, b) = \frac{1}{\sigma\sqrt{2\pi}} \int_a^b e^{-(x-x_0)^2/2\sigma^2}. \tag{6.93}$$

This cannot, in general, be expressed as a simple function. As in the case of the gamma function in the previous section, the *error function* can be defined by a definite integral

$$\operatorname{erf} x \equiv \frac{2}{\sqrt{\pi}} \int_0^x e^{-t^2}\, dt. \tag{6.94}$$

The constant $2/\sqrt{\pi}$ is chosen so that $\operatorname{erf}(\infty) = 1$. Note also that $\operatorname{erf}(0) = 0$ and that erf is an odd function. It is also useful to define the *complementary error function*

$$\operatorname{erfc} x \equiv 1 - \operatorname{erf} x = \frac{2}{\sqrt{\pi}} \int_x^{\infty} e^{-t^2}\, dt. \tag{6.95}$$

These functions are graphed in Figure 6.10. The integral (6.93) reduces to

$$P(a, b) = \frac{1}{2}\left[\operatorname{erf}\left(\frac{x_0 - a}{\sqrt{2}\sigma}\right) - \operatorname{erf}\left(\frac{x_0 - b}{\sqrt{2}\sigma}\right) \right]. \tag{6.96}$$

In particular, the fraction of a Gaussian distribution beyond one standard deviation on either side is given by

$$P(x_0 + \sigma, \infty) - P(-\infty, x_0 - \sigma) = \frac{1}{2}\operatorname{erfc}\left(\frac{1}{\sqrt{2}}\right) \approx 0.1587. \tag{6.97}$$

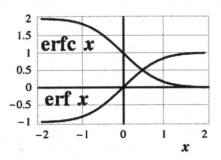

Figure 6.10 Error function erf x and complementary error function erfc x. The curve for erf x closely resembles arctan x.

This means that about 68.3% of the probability lies between $x_0 \pm \sigma$.

The average IQ of college graduates has been estimated to lie in the range 114–115, about one sigma above the average for the population as a whole. College professors allegedly have an average IQ around 132. Thus the chance is only about 15% that you are smarter than your Professor. But, although you can't usually best him or her by raw brainpower, you can still do very well with "street smarts" which you are hopefully acquiring from this book.

Problem 6.10.1. Evaluate the definite integral $\int_1^2 e^{-x^2}\, dx$. You will need to look up or compute values of the error function.

6.11 Numerical Integration

It may be difficult or even impossible in some cases to express an indefinite integral $\int f(x)\, dx$ in analytic form. Or the function $f(x)$ might be in the form of numerical data with $f(x_n) = y_n$. One can still find accurate numerical approximations for corresponding definite integrals $\int_a^b f(x)\, dx$. The most elementary method is very nearly a restatement of the definition of a Riemann sum, Eq. 6.14, but uses a row of trapezoidal strips to approximate the integral, as shown in Figure 6.11. According to the *trapezoidal rule* the integral is approximated by the sum of the areas of the pink trapezoids. Using the notation $a = x_0$, $b = x_4$, this can be written

$$\int_a^b f(x)dx \approx \frac{1}{2}\Delta x(y_0 + y_1) + \frac{1}{2}\Delta x(y_1 + fy_2) + \frac{1}{2}\Delta x(y_2 + y_3)$$

$$+ \frac{1}{2}\Delta x(y_3 + y_4) = \Delta x\left(\frac{1}{2}y_0 + y_1 + y_2 + y_3 + \frac{1}{2}y_4\right). \quad (6.98)$$

More generally, approximating the integral using n trapezoids,

$$\int_a^b f(x)dx \approx \Delta x\left(\frac{1}{2}y_0 + y_1 + y_2 + y_3 + \cdots + \frac{1}{2}y_n\right). \quad (6.99)$$

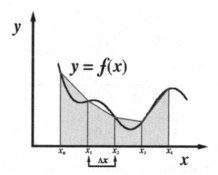

Figure 6.11 Trapezoidal rule for numerical integration.

Clearly, the result will become more accurate as n is increased and Δx is decreased.

A somewhat better numerical approximation can be obtained using *Simpson's rule*. This requires an even value of n and can be expressed as a summation analogous to Eq. (6.99), but with modified values of the coefficients:

$$\int_a^b f(x)dx \approx \frac{\Delta x}{3}\left(y_0 + 4y_1 + 2y_2 + 4y_3 + 2y_4 + \cdots + y_n\right). \tag{6.100}$$

Note that the first and last terms have coefficients 1, while the intermediate coefficients alternate between 4 and 2. The basis of Simpson's rule is the replacement of the linear segments atop the trapezoidal strips by parabolic arcs over each pair of strips. The three y values associated with two adjacent strips can be fitted to a parabola $y = ax^2 + bx + c$. Integration then gives

$$\int_{x_0}^{x_2} y(x)\, dx \approx \frac{\Delta x}{6}(y_0 + 4y_1 + y_2) \tag{6.101}$$

and summation over $n/2$ of such adjacent structures leads to Simpson's rule.

There exist even more accurate formulas for numerical integration, but it is more convenient to turn the work over to a computer. In Mathematica, the command `NIntegrate[f[x],{x,a,b}]`, with a specified function $f(x)$ and limits a and b, performs the integration using an appropriate algorithm, if necessary, with recursively varying subdivisions.

Problem 6.11.1. Evaluate the integral $\int_1^2 e^{-x^2} dx$ using the trapezoidal rule and Simpson's rule. Try subdivisions with $n =$ and $n = 6$ for each. Compare with the exact value calculated in Problem 6.10.1.

7 Series and Integrals

When you press the SIN or LOG key on your scientific calculator, it almost instantly returns a numerical value. What is really happening is that the microprocessor inside the calculator is summing a series representation of that function, similar to the series we have already encountered for $\sin x$, $\cos x$, and e^x in Chapters 3 and 4. Power series (sums of powers of x) and other series of functions are very important mathematical tools, both for computation and for deriving theoretical results.

7.1 Some Elementary Series

An *arithmetic progression* is a sequence such as $1, 4, 7, 10, 13, 16$. As shown in Section 1.2, the sum of an arithmetic progression with n terms is given by

$$S_n = a_0 + a_1 + \cdots + a_{n-1} = \sum_{k=0}^{n-1} a_k = \frac{n}{2}(a + \ell),$$

$$a_n = a_0 + nd, \quad a_{n-1} = a_0 + (n-1)d, \tag{7.1}$$

where $a = a_0$ is the first term, d is the constant difference between terms, and $\ell = a_{n-1}$ is the last term.

A *geometric progression* is a sequence which increases or decreases by a common factor r, for example, $1, 3, 9, 27, 81, \ldots$ or $1, 1/2, 1/4, 1/8, 1/16, \ldots$ The sum of a geometric progression is given by,

$$S_n = a + ar + ar^2 + ar^3 + \cdots + ar^{n-1} = \sum_{k=0}^{n-1} ar^k = \frac{a - ar^{n-1}}{1 - r}. \tag{7.2}$$

When $r < 1$, the sum can be carried to infinity to give

$$S_\infty = \sum_{k=0}^{\infty} ar^k = \frac{a}{1 - r}. \tag{7.3}$$

We had already found from the binomial theorem that

$$\frac{1}{1 - x} = 1 + x + x^2 + x^3 + \cdots. \tag{7.4}$$

Guide to Essential Math 2e. http://dx.doi.org/10.1016/B978-0-12-407163-6.00007-2

By applying the binomial theorem successively to $[1 - x(x + 1)]^{-1}$, then to each $[x(x + 1)]^n$ you can show that

$$f(x) = \frac{1}{1 - x(x + 1)} = \frac{1}{1 - x - x^2}$$

$$= 1 + x + 2x^2 + 3x^3 + 5x^4 + 8x^5 + 13x^6 + 21x^7 + 34x^8 + \cdots . \quad (7.5)$$

Therefore $f(x)$ serves as a *generating function* for the Fibonacci numbers:

$$f(x) = \sum_{n=0}^{\infty} F_n x^n. \quad (7.6)$$

An infinite geometric series inspired by one of Zeno's paradoxes is

$$\frac{1}{2} + \frac{1}{4} + \frac{1}{8} + \frac{1}{16} + \cdots = 1. \quad (7.7)$$

Zeno's paradox of motion claims that if you shoot an arrow, it can never reach its target. First it has to travel half way, then half way again—meaning 1/4 of the distance—then continue with an *infinite* number of steps, each taking it $1/2^n$ closer. Since infinity is so large, you'll never get there! What we now understand that Zeno possibly didn't (some scholars believe that his argument was meant to be satirical) was that an infinite number of decreasing terms can add up to a finite quantity.

The integers 1 to to n add up to

$$\sum_{k=1}^{n} k = \frac{n(n + 1)}{2}, \quad (7.8)$$

while the sum of the squares is given by

$$\sum_{k=1}^{n} k^2 = \frac{n(n + 1)(2n + 1)}{6} \quad (7.9)$$

and the sum of the cubes by

$$\sum_{k=1}^{n} k^3 = \left[\frac{n(n + 1)}{2} \right]^2 . \quad (7.10)$$

7.2 Power Series

Almost all functions can be represented by *power series* of the form

$$f(x) = \alpha(x) \sum_{n=0}^{\infty} a_n (x - x_0)^n, \quad (7.11)$$

where $\alpha(x)$ might be a factor such as x^{α}, $e^{-\alpha x^2}$, or $\ln x$. The case $\alpha(x) = 1$ and $x_0 = 0$ provides the most straightforward class of power series. We are already familiar with the series for the exponential function:

$$e^x = 1 + x + \frac{x^2}{2!} + \frac{x^3}{3!} + \cdots \tag{7.12}$$

as well as the sine and cosine:

$$\sin x = x - \frac{x^3}{3!} + \frac{x^5}{5!} - \frac{x^7}{7!} + \cdots , \tag{7.13}$$

$$\cos x = 1 - \frac{x^2}{2!} + \frac{x^4}{4!} - \frac{x^6}{6!} + \cdots . \tag{7.14}$$

Recall also the binomial expansion:

$$(1 + x)^n = 1 + nx + \frac{n(n - 1)}{2} x^2 + \cdots . \tag{7.15}$$

Useful results can be obtained when power series are differentiated or integrated term by term. This is a valid procedure under very general conditions.

Consider, for example, the binomial expansion

$$(1 + x^2)^{-1} = 1 - x^2 + x^4 - x^6 + \cdots . \tag{7.16}$$

Making use of the known integral

$$\int \frac{dx}{1 + x^2} = \arctan x, \tag{7.17}$$

we obtain a series for the arctangent

$$\arctan x = \int (1 - x^2 + x^4 - \cdots) \, dx = x - \frac{x^3}{3} + \frac{x^5}{5} - \cdots = \sum_{n=0}^{\infty} \frac{x^{2n-1}}{2n - 1}. \tag{7.18}$$

With $\arctan 1 = \pi/4$, this gives a famous series for π

$$\frac{\pi}{4} = 1 - \frac{1}{3} + \frac{1}{5} - \frac{1}{7} + \cdots \tag{7.19}$$

usually attributed to Gregory and Leibniz.

A second example begins with another binomial expansion

$$(1 + x)^{-1} = 1 - x + x^2 - x^3 + \cdots . \tag{7.20}$$

We can again evaluate the integral

$$\int \frac{dx}{1 + x} \overset{y=1+x}{\longrightarrow} \int \frac{dy}{y} = \ln y = \ln(1 + x). \tag{7.21}$$

This gives a series representation for the natural logarithm:

$$\ln(1+x) = x - \frac{x^2}{2} + \frac{x^3}{3} - \frac{x^4}{4} + \cdots . \tag{7.22}$$

For $x = 1$, this gives another famous series

$$\ln 2 = 1 - \frac{1}{2} + \frac{1}{3} - \frac{1}{4} + \cdots . \tag{7.23}$$

As a practical matter, the convergence of this series is excruciatingly slow. It takes about 1000 terms to get $\ln 2$ correct to three significant figures, 0.693.

It is relatively simple to multiply two power series. Note that

$$\left(\sum_{n=0}^{\infty} a_n x^n \right) \times \left(\sum_{n=0}^{\infty} b_n x^n \right) = \sum_{n=0}^{\infty} \sum_{m=0}^{\infty} a_n b_m x^{n+m} = \sum_{n=0}^{\infty} \sum_{k=0}^{n} a_n b_{n-k} x^n . \tag{7.24}$$

Problem 7.2.1. Evaluate the first few terms of the power series for $\sin^2 x$ and $\cos x \sin x$.

Inversion of a power series means, in principle, solving for the expansion variable x as a function of the sum of the series. Consider the special case of a series in the form

$$y = x + a_2 x^2 + a_3 x^3 + a_4 x^4 + a_5 x^5 + \cdots . \tag{7.25}$$

The constants a_0 and a_1 in the more general case can be absorbed into the variable y and the remaining constants. Then x can be expanded in powers of y in the form

$$x = y + b_2 y^2 + b_3 y^3 + b_4 y^4 + b_5 y^5 + \cdots . \tag{7.26}$$

By a recursive procedure, carried out only as far as $n = 5$, we obtain:

$$\begin{aligned} &b_2 = -a_2, \quad b_3 = 2a_2^2 - a_3, \quad b_4 = 5a_2 a_3 - a_4 - 5a_2^3, \\ &b_5 = 6a_2 a_4 + 3a_3^2 + 14a_2^4 - a_5 - 21a_2^2 a_3, \ldots \end{aligned} \tag{7.27}$$

Problem 7.2.2. Invert the power series for $\sin x$ to find the first few terms in the power series for $\arcsin x$.

7.3 Convergence of Series

The *partial sum* S_n of an infinite series is the sum of the first n terms:

$$S_n \equiv a_1 + a_2 + \cdots + a_n = \sum_{k=1}^{n} a_k . \tag{7.28}$$

The series is *convergent* if

$$\lim_{n \to \infty} S_n = S, \tag{7.29}$$

where S is a finite quantity. A *necessary* condition for convergence, thus a preliminary test, is that

$$\lim_{n\to\infty} a_n = 0. \tag{7.30}$$

Several tests for convergence that are usually covered in introductory calculus courses. The *comparison test*: if a series $\sum_{n=1}^{\infty} b_n$ is known to converge and $a_n \leqslant b_n$ for all n, then the series $\sum_{n=1}^{\infty} a_n$ is also convergent. The *ratio test*: if $\lim_{n\to\infty} a_{n+1}/a_n < 1$ then the series converges. There are more sensitive ratio tests in the case that the limit approaches 1, but you will rarely need these outside of math courses. The most useful test for convergence is the *integral test*. This is based on turning things around using our original definition of an integral as the limit of a sum. The sum $\sum_{n=1}^{\infty} a_n$ can be approximated by an integral by turning the discrete variable n into a continuous variable x. If the integral

$$\int_1^{\infty} a(x)dx \tag{7.31}$$

is finite, then the original series converges.

A general result is that any decreasing alternating series, such as (7.23), converges. *Alternating* refers to the alternation of plus and minus signs. How about the analogous series with all plus signs?:

$$1 + \frac{1}{2} + \frac{1}{3} + \frac{1}{4} + \cdots . \tag{7.32}$$

After 1000 terms the sum equals 7.485. It might appear that, with sufficient patience, the series will eventually converge to a finite quantity. Not so, however! The series is divergent and sums to infinity. This can be seen by applying the integral test:

$$\sum_{n=1}^{\infty} \frac{1}{n} \approx \int_1^{\infty} \frac{dx}{x} = \ln\infty = \infty. \tag{7.33}$$

A finite series of the form

$$H_n = \sum_{k=1}^{n} \frac{1}{k} \tag{7.34}$$

is called a *harmonic series*. Using the same approximation by an integral, we estimate that this sum is approximately equal to $\ln n$. The difference between H_n and $\ln n$ was shown by Euler to approach a constant as $n \to \infty$:

$$\lim_{n\to\infty} \sum_{k=1}^{n} \frac{1}{k} - \ln n \equiv \gamma \approx 0.5772, \tag{7.35}$$

where γ (sometimes denoted C) is known as the *Euler-Mascheroni constant*. It comes up frequently in mathematics, for example, the integral

$$\int_0^{\infty} e^{-x} \ln x \, dx = -\gamma. \tag{7.36}$$

An alternating series is said to be *absolutely convergent* if the corresponding sum of absolute values, $\sum_{n=1}^{\infty} |a_n|$, is also convergent. This is *not* true for the alternating harmonic series (7.23). Such a series is said to be *conditionally convergent*. Conditionally convergent series must be treated with extreme caution. A theorem due to Riemann states that, by a suitable rearrangement of terms, a conditionally convergent series may be made to equal any desired value, or to diverge. Consider, for example, the series (7.20) for $1/(1+x)$ when $x = 1$:

$$\frac{1}{1+1} \stackrel{?}{=} 1 - 1 + 1 - 1 + 1 - \cdots .$$
(7.37)

Different ways of grouping the terms of the series give different answers. Thus $(1 - 1) + (1 - 1) + (1 - 1) + \cdots = 0$, while $1 + (-1 + 1) + (-1 + 1) + (1 - 1) + \cdots = 1$. But $1/(1 + 1)$ actually equals 1/2. Be very careful!

7.4 Taylor Series

There is a systematic procedure for deriving power-series expansions for all well-behaved functions. Assuming that a function $f(x)$ can be represented in a power series about $x_0 = 0$, we write

$$f(x) = a_0 + a_1 x + a_2 x^2 + a_3 x^3 + \cdots = \sum_{n=0}^{\infty} a_n x^n.$$
(7.38)

Clearly, when $x = 0$,

$$f(0) = a_0.$$
(7.39)

The first derivative of $f(x)$ is given by

$$f'(x) = a_1 + 2a_2 x + 3a_3 x^2 + \cdots = \sum_{n=0}^{\infty} n a_n x^{n-1}$$
(7.40)

and, setting $x = 0$, we obtain

$$f'(0) = a_1.$$
(7.41)

The second derivative is given by

$$f''(x) = 2a_2 + 3 \cdot 2a_3 x + 4 \cdot 3a_4 x^2 + \cdots = \sum_{n=0}^{\infty} n(n-1) a_n x^{n-2}$$
(7.42)

with

$$f''(0) = 2a_2.$$
(7.43)

With repeated differentiation, we find

$$f^{(n)}(0) = n!a_n, \tag{7.44}$$

where

$$f^{(n)}(0) = \frac{d^n}{dx^n} f(x)\Big|_{x=0}. \tag{7.45}$$

We have therefore determined the coefficients $a_0, a_1, a_2, \cdots$ in terms of derivatives of $f(x)$ evaluated at $x = 0$ and the expansion (7.38) can be given more explicitly by

$$f(x) = f(0) + xf'(0) + \frac{x^2}{2} f''(0) + \frac{x^3}{3!} f'(0) + \cdots$$

$$= \sum_{n=0}^{\infty} \frac{x^n}{n!} f^{(n)}(0). \tag{7.46}$$

If the expansion is carried out around $x_0 = a$, rather than 0, the result generalizes to

$$f(x) = f(a) + (x - a)f'(a) + \frac{(x - a)^2}{2} f''(a)$$

$$+ \frac{(x - a)^3}{3!} f'''(a) + \cdots = \sum_{n=0}^{\infty} \frac{(x - a)^n}{n!} f^{(n)}(a). \tag{7.47}$$

This result is known as *Taylor's theorem* and the expansion is a *Taylor series*. The case $x_0 = 0$, given by Eq. (7.46), is sometimes called a *Maclaurin series*.

In order for a Taylor series around $x = a$ to be valid it is necessary for all derivatives $f^{(n)}(a)$ to exist. The function is then said to be *analytic* at $x = a$. A function which is not analytic at one point can still be analytic at other points. For example, $\ln x$ is not analytic at $x = 0$ but *is* at $x = 1$. The series (7.22) is equivalent to an expansion of $\ln x$ around $x = 1$.

We can now systematically derive all the series we obtained previously by various other methods. For example, given $f(x) = (1 + x)^\alpha$, we find

$$f^{(n)}(0) = \alpha(\alpha - 1)(\alpha - 2) \cdots (\alpha - n + 1) = \frac{\alpha!}{(\alpha - n)!} \tag{7.48}$$

so that (7.47) gives

$$(1 + x)^\alpha = \sum_{n=0}^{\infty} \frac{\alpha!}{(\alpha - n)!n!} x^n, \tag{7.49}$$

which is the binomial expansion. This result is seen to be valid even for noninteger values of α. In the latter case we should express (7.49) in terms of the gamma function as follows:

$$(1 + x)^\alpha = \sum_{n=0}^{\infty} \frac{\Gamma(\alpha + 1)}{\Gamma(\alpha - n + 1)n!} x^n. \tag{7.50}$$

The series for $f(x) = e^x$ is easy to derive because $f^{(n)}(x) = e^x$ for all n. Therefore, as we have already found

$$e^x = \sum_{n=0}^{\infty} \frac{x^n}{n!}. \tag{7.51}$$

The Taylor series for $\sin x$ is also straightforward since successive derivatives cycle among $\sin x$, $-\cos x$, $-\sin x$, and $\cos x$. Since $\sin 0 = 0$ and $\cos 0 = 1$, the series expansion contains only odd powers of x with alternating signs:

$$\sin x = x - \frac{x^3}{3!} + \frac{x^5}{5!} - \cdots. \tag{7.52}$$

Analogously, the expansion for the cosine is given by

$$\cos x = 1 - \frac{x^2}{2!} + \frac{x^4}{4!} - \cdots. \tag{7.53}$$

Euler's theorem $e^{\pm ix} = \cos x \pm i \sin x$ can then be deduced by comparing the series for these three functions.

Problem 7.4.1. Work out the Taylor series for the function $f(x) = x/e^x - 1$ about $x = 0$, up to the term in x^4. This is a little tricky since you have to evaluate limits as $x \to 0$ to determine the derivatives (0).

7.5 Bernoulli and Euler Numbers

The answer to Problem 7.4.1 is the series

$$\frac{x}{(e^x - 1)} = 1 - \frac{1}{2}x + \frac{1}{12}x^2 - \frac{1}{720}x^4 + \frac{1}{30240}x^6 + \cdots. \tag{7.54}$$

This series provides a generating function for the *Bernoulli numbers*, B_n, whereby

$$\frac{x}{(e^x - 1)} = \sum_{n=0}^{\infty} \frac{B_n x^n}{n!} \quad |x| < 2\pi. \tag{7.55}$$

Explicitly,

$$B_0 = 1, \quad B_1 = -\frac{1}{2}, \quad B_2 = \frac{1}{6}, \quad B_4 = -\frac{1}{30}, \quad B_6 = \frac{1}{42},$$
$$B_8 = \frac{1}{30}, \quad B_{10} = \frac{5}{66}, \cdots$$
$$B_3 = B_5 = B_7 = B_9 = \cdots = 0. \tag{7.56}$$

Bernoulli numbers find application in number theory, in numerical analysis, and in expansions of several functions related to $\tan x$ and $\tanh x$.

A symbolic relation which can be used to evaluate Bernoulli numbers mimics the binomial expansion:

$$B^n = (1 + B)^n, \qquad n \geqslant 2, \tag{7.57}$$

where B^n is to be replaced by B_n.

We can obtain an expansion for $\coth(x/2)$ in the following steps involving the Bernoulli numbers:

$$\frac{x}{e^x - 1} + \frac{x}{2} = 1 + \sum_{n=2}^{\infty} \frac{B_n x^n}{n!}. \tag{7.58}$$

But

$$\frac{x}{e^x - 1} + \frac{x}{2} = \frac{x}{2} \times \frac{e^x + 1}{e^x - 1} = \frac{x}{2} \times \frac{e^{x/2} + e^{-x/2}}{e^{x/2} - e^{-x/2}} = \frac{x}{2} \coth\left(\frac{x}{2}\right). \tag{7.59}$$

Therefore

$$\frac{x}{2} \coth\left(\frac{x}{2}\right) = 1 + \sum_{n=2}^{\infty} \frac{B_n x^n}{n!} \tag{7.60}$$

and, more explicitly,

$$\coth\left(\frac{x}{2}\right) = \frac{x}{2} - \frac{x^3}{24} + \frac{x^4}{240} - \frac{17 x^7}{40320} + \cdots. \tag{7.61}$$

Problem 7.5.1. Find the expansions for $\coth x$ and for $\cot x$.

Problem 7.5.2. From the trigonometric identity $\tan x = \cot x - \cot 2x$, work out the expansions for $\tan x$.

Somewhat analogous to the definition of Bernoulli numbers are the *Euler numbers* E_n. These can be obtained from the generating function:

$$\frac{2e^x}{e^{2x} + 1} = \sum_{n=0}^{\infty} \frac{E_n x^n}{n!}, \qquad |x| < \frac{\pi}{2}. \tag{7.62}$$

The first few Euler numbers are

$$E_0 = 1, \quad E_2 = -1, \quad E_4 = 51, \quad E_6 = -611, \quad E_8 = 13851, \ldots \tag{7.63}$$

with all equal to 0 for odd indices n. We find directly that

$$\frac{2e^x}{e^{2x} + 1} = \frac{2}{e^x + e^{-x}} = \operatorname{sech} x \tag{7.64}$$

and thus obtain the expansion

$$\operatorname{sech} x = 1 - \frac{x^2}{2} + \frac{5 x^4}{24} - \frac{61 x^6}{720} + \frac{277 x^8}{8064} + \cdots. \tag{7.65}$$

Problem 7.5.3.   Find the corresponding expansion for $\sec x$.

The Euler-Maclaurin sum formula provides a powerful method for evaluating some difficult summations. It actually represents a more precise connection between Riemann sums and their corresponding integrals. It can be used to approximate finite sums and even infinite series using integrals with some additional terms involving Bernoulli numbers. It can be shown that

$$\sum_{n=a}^{b} f(n) \approx \int_a^b f(x)\,dx + \frac{f(b) - f(a)}{2} + \sum_{n=2}^{n_{max}} \frac{B_n}{n!}\left[f^{(n-1)}(b) - f^{(n-1)}(a)\right].$$
(7.66)

Problem 7.5.4.   Using the Euler-Maclaurin formula, evaluate the sum $\sum_{k=1}^{n} k^3$.

7.6   L'Hôpital's Rule

The value of a function is called an *indeterminate form* at some point a if its limit as $x \to a$ apparently approaches one of the forms $0/0$, ∞/∞, or $0\cdot\infty$. Two examples are the combinations $\sin x/x$ and $x\ln x$ as $x \to 0$. (As we used to say in high school, such sick functions had to be sent to L'Hôspital to be cured.) To be specific, let us consider a case for which

$$\lim_{x\to 0} \frac{f(x)}{g(x)} = \frac{0}{0}.$$
(7.67)

If $f(x)$ and $g(x)$ are both expressed in Taylor series about $x = 0$,

$$\frac{f(x)}{g(x)} = \frac{f(0) + xf'(0) + (x^2/2)f''(0) + \cdots}{g(0) + xg'(0) + (x^2/2)g''(0) + \cdots}.$$
(7.68)

If $f(0)$ and $g(0)$ both equal 0 but $f'(0)$ and $g'(0)$ are finite, the limit in (7.67) is given by

$$\lim_{x\to 0} \frac{f(x)}{g(x)} = \frac{f'(0)}{g'(0)},$$
(7.69)

a result known as *L'Hôpital's rule*. In the event that one or both first derivatives also vanishes, the lowest order nonvanishing derivatives in the numerator and denominator determine the limit.

To evaluate the limit of $\sin x/x$, let $f(x) = \sin x$ and $g(x) = x$. In this case, $f(0) = g(0) = 0$. But $f'(x) = \cos x$, so $f'(0) = 1$. Also, $g'(x) = 1$, for all x. Therefore

$$\lim_{x\to 0} \frac{\sin x}{x} = 1.$$
(7.70)

This is also consistent with the approximation that $\sin x \approx x$ for $x \to 0$. For the limit of $x\ln x$, let $f(x) = \ln x$, $g(x) = 1/x$. Now $f'(x) = 1/x$ and $g'(x) = -1/x^2$. We find therefore that

$$\lim_{x\to 0} x\ln x = 0.$$
(7.71)

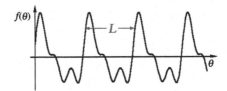

Figure 7.1 Periodic function $f(\theta)$ with period L.

Problem 7.6.1. Evaluate the limit

$$\lim_{x \to 0} \frac{e^x - 1 - x}{x^2}.$$

7.7 Fourier Series

Taylor series expansions, as described above, provide a very general method for representing a large class of mathematical functions. For the special case of *periodic functions*, a powerful alternative method is expansion in an infinite sum of sines and cosines, known as a *trigonometric series* or *Fourier series*. A periodic function is one which repeats in value when its argument is increased by multiples of a constant L, called the *period* or *wavelength*. For example,

$$f(\theta + kL) = f(\theta), \quad k = 0, \pm 1, \pm 2, \ldots \tag{7.72}$$

as shown in Figure 7.1. For convenience, let us consider the case $L = 2\pi$. Sine and cosine are definitive examples of functions periodic in 2π since

$$\sin(\theta + 2k\pi) = \sin\theta \quad \text{and} \quad \cos(\theta + 2k\pi) = \cos\theta. \tag{7.73}$$

The functions $\sin n\theta$ and $\cos n\theta$, with n=integer, are also periodic in 2π (as well as in $2\pi/n$).

A function $f(\theta)$ periodic in 2π can be expanded as follows:

$$f(\theta) = \frac{a_0}{2} + a_1 \cos\theta + a_2 \cos 2\theta + \cdots + b_1 \sin\theta + b_2 \sin 2\theta + \cdots$$

$$= \frac{a_0}{2} + \sum_{n=1}^{\infty} (a_n \cos n\theta + b_n \sin n\theta). \tag{7.74}$$

Writing the constant term as $a_0/2$ is convenient, as will be seen shortly. The coefficients a_n, b_n can be determined by making use of the following definite integrals:

$$\int_0^{2\pi} \cos n\theta \cos m\theta \, d\theta = \pi \delta_{nm}, \tag{7.75}$$

$$\int_0^{2\pi} \sin n\theta \sin m\theta \, d\theta = \pi \delta_{nm}, \tag{7.76}$$

$$\int_0^{2\pi} \cos n\theta \sin m\theta \, d\theta = 0 \quad \text{all } m, n. \tag{7.77}$$

These integrals are expressed compactly with use of the *Kronecker delta*, defined as follows:

$$\delta_{nm} \equiv \begin{cases} 0 & \text{if } m \neq n, \\ 1 & \text{if } m = n. \end{cases} \tag{7.78}$$

Two functions are said to be *orthogonal* if the definite integral of their product equals zero. (Analogously, two vectors $\mathbf{a}$ and $\mathbf{b}$ whose scalar product $\mathbf{a} \cdot \mathbf{b}$ equals zero are said to be orthogonal—meaning perpendicular, in that context.) The set of functions $\{1, \sin\theta, \cos\theta, \sin 2\theta, \cos 2\theta, \ldots\}$ is termed an *orthogonal set* in an interval of width $2n\pi$.

The nonvanishing integrals in (7.75) and (7.76) follow easily from the fact that $\sin^2\theta$ and $\cos^2\theta$ have average values of $\frac{1}{2}$ over an integral number of wavelengths. Thus

$$\int_0^{2\pi} \sin^2 n\theta \, d\theta = \int_0^{2\pi} \cos^2 n\theta \, d\theta = \frac{1}{2} \times 2\pi = \pi. \tag{7.79}$$

The orthogonality relations (7.75)–(7.77) enable us to determine the Fourier expansion coefficients a_n and b_n. Consider the integral $\int_0^{2\pi} f(\theta)\cos n\theta \, d\theta$, with $f(\theta)$ expanded using (7.74). By virtue of the orthogonality relations, only one term in the expansion survives integration:

$$\int_0^{2\pi} f(\theta)\cos n\theta \, d\theta = \cdots + 0 + a_n \int_0^{2\pi} \cos n\theta \cos n\theta \, d\theta + 0 + \cdots. \tag{7.80}$$

Solving for a_n we find

$$a_n = \frac{1}{\pi} \int_0^{2\pi} f(\theta)\cos n\theta \, d\theta, \qquad n = 0, 1, 2, \ldots \tag{7.81}$$

Note that the case $n = 0$ correctly determines a_0, by virtue of the factor $\frac{1}{2}$ in Eq. (7.74). Analogously, the coefficients b_n are given by

$$b_n = \frac{1}{\pi} \int_0^{2\pi} f(\theta)\sin n\theta \, d\theta, \qquad n = 1, 2, 3, \ldots \tag{7.82}$$

In about half of textbooks, the limits of integration in Eqs. (7.81) and (7.82) are chosen as $-\pi \leqslant \theta \leqslant \pi$ rather than $0 \leqslant \theta \leqslant 2\pi$. This is just another way to specify one period of the function and the same results are obtained in either case.

As an illustration, let us calculate the Fourier expansion for a *square wave*, defined as follows:

$$f(\theta) = \begin{cases} +1 & \text{for } 0 \leqslant \theta \leqslant \pi, 2\pi \leqslant \theta \leqslant 3\pi, \ldots \\ -1 & \text{for } \pi \leqslant \theta \leqslant 2\pi, 3\pi \leqslant \theta \leqslant 4\pi, \ldots \end{cases} \tag{7.83}$$

A square-wave oscillator is often used to test the frequency response of an electronic circuit. The Fourier coefficients can be computed using (7.81) and (7.82). First we find

$$a_n = \frac{1}{\pi} \int_0^{2\pi} f(\theta)\cos n\theta \, d\theta = \frac{1}{\pi} \int_0^{\pi} \cos n\theta \, d\theta - \frac{1}{\pi} \int_\pi^{2\pi} \cos n\theta \, d\theta = 0. \tag{7.84}$$

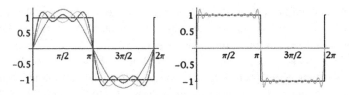

Figure 7.2 Fourier series approximating square wave $f(\theta)$. Partial sums S_1, S_2, S_3, and S_{10} are shown.

Thus all the cosine contributions equal zero since $\cos n\theta$ is symmetrical about $\theta = \pi$. The square wave is evidently a Fourier sine series, with only nonvanishing b_n coefficients. We find

$$b_n = \frac{1}{\pi}\int_0^{2\pi} f(\theta)\sin n\theta\, d\theta = \frac{1}{\pi}\int_0^{\pi}\sin n\theta\, d\theta - \frac{1}{\pi}\int_{\pi}^{2\pi}\sin n\theta\, d\theta \qquad (7.85)$$

giving

$$b_n = 0 \quad \text{for } n \text{ even}, \quad b_n = \frac{4}{n\pi} \quad \text{for } n \text{ odd}. \qquad (7.86)$$

The Fourier expansion for a square wave can thus be written

$$\begin{aligned} f(\theta) &= \frac{4}{\pi}\sin\theta + \frac{4}{3\pi}\sin 3\theta + \frac{4}{5\pi}\sin 5\theta + \cdots \\ &= \sum_{n=1}^{\infty}\frac{4}{(2n-1)\pi}\sin(2n-1)\theta. \end{aligned} \qquad (7.87)$$

For $\theta = \pi$, Eq. (7.87) reduces to (7.19), the Gregory-Leibniz series for $\pi/4$:

$$\frac{\pi}{4} = 1 - \frac{1}{3} + \frac{1}{5} - \frac{1}{7} + \cdots. \qquad (7.88)$$

A Fourier series carried through N terms is called a *partial sum* S_N. Figure 7.2 shows the function $f(\theta)$ and the partial sums S_1, S_2, S_3, and S_{10}. Note how the higher partial sums overshoot near the points of discontinuity $\theta = 0, \pi, 2\pi$. This is known as the *Gibbs phenomenon*. As $N \to \infty$, $S_N \approx \pm 1.18$ at these points.

Problem 7.7.1. Calculate the first several terms of the Fourier series representing a sawtooth signal, shown below.

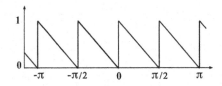

The general conditions for a periodic function $f(\theta)$ to be representable by a Fourier series are the following. The function must have a finite number of maxima and minima and a finite number of discontinuities between 0 and 2π. Also, $\int_0^{2\pi} |f(\theta)|^2 \, d\theta$ must be finite. If these conditions are fulfilled then the Fourier series (7.74) with coefficients given by (7.81) and (7.82) converges to $f(\theta)$ at points where the function is continuous. At discontinuities θ_0, the Fourier series converges to the midpoint of the jump, $[f(\theta_0^+) + f(\theta_0^-)]/2$.

Recall that sines and cosines can be expressed in terms of complex exponential functions, according to Eqs. (4.60) and (4.61). Accordingly, a Fourier series can be expressed in a more compact form:

$$f(\theta) = \sum_{m=-\infty}^{\infty} c_m e^{im\theta}, \tag{7.89}$$

where the coefficients c_m might be complex numbers. The orthogonality relations for complex exponentials are given by

$$\int_0^{2\pi} (e^{im'\theta})^* e^{im\theta} \, d\theta = \int_0^{2\pi} e^{i(m-m')\theta} \, d\theta = 2\pi \, \delta_{mm'}. \tag{7.90}$$

These determine the complex Fourier coefficients:

$$c_m = \frac{1}{2\pi} \int_0^{2\pi} f(\theta) e^{-im\theta} \, d\theta, \qquad m = 0, \pm 1, \pm 2, \ldots \tag{7.91}$$

If $f(\theta)$ is a real function, then $c_{-m} = c_m{}^*$ for all m.

For functions with a periodicity L different from 2π, the variable θ can be replaced by $2\pi x/L$. The formulas for Fourier series are then modified as follows:

$$f(x) = \frac{a_0}{2} + \sum_{n=1}^{\infty} \left[a_n \cos\left(\frac{2n\pi x}{L}\right) + b_n \sin\left(\frac{2n\pi x}{L}\right) \right] \tag{7.92}$$

with

$$a_n = \frac{2}{L} \int_0^L f(x) \cos\left(\frac{2n\pi x}{L}\right) dx \tag{7.93}$$

and

$$b_n = \frac{2}{L} \int_0^L f(x) \sin\left(\frac{2n\pi x}{L}\right) dx. \tag{7.94}$$

For complex Fourier series,

$$f(x) = \sum_{m=-\infty}^{\infty} c_m \, e^{2\pi i m x/L} \tag{7.95}$$

with

$$c_m = \frac{1}{L} \int_0^L f(x) e^{-2\pi i m x/L} \, dx = \frac{1}{L} \int_{-L/2}^{L/2} f(x) e^{-2\pi i m x/L} \, dx. \tag{7.96}$$

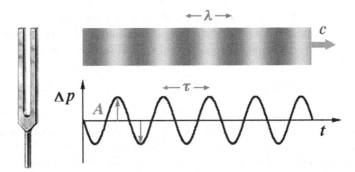

Figure 7.3 Sound wave of a single frequency produced by a tuning fork. The upper part of the figure shows an instantaneous view of a longitudinal sound wave traveling at speed $c \approx 350$ m/s. Wavelength and period are related by $\lambda/\tau = c$.

Many of the applications of Fourier analysis involve the time-frequency domain. A time-dependent signal $f(t)$ can be expressed

$$f(t) = \frac{a_0}{2} + \sum_{n=1}^{\infty} (a_n \cos \omega_n t + b_n \sin \omega_n t), \tag{7.97}$$

where the ω_n are frequencies expressed in radians per second.

When a tuning fork is struck, it emits a tone which can be represented by a sinusoidal wave—one having the shape of a sine or cosine. For tuning musical instruments, a fork might be machined to produce a pure tone at 440 Hz, which corresponds to A above middle C. (Middle C would then have a frequency of 278.4375 Hz.) The graph in Figure 7.3 shows the variation of air pressure (or density) with time for a sound wave, as measured at a single point. Δp represents the deviation from the undisturbed atmospheric pressure p_0. The maximum variation of Δp above or below p_0 is called the *amplitude* A of the wave. The time between successive maxima of the wave is called the *period* τ. Since the argument of the sine or cosine varies between 0 and 2π in one period, the form of the wave could be the function

$$\psi(t) = A \sin\left(\frac{2\pi t}{\tau}\right) = A \sin 2\pi \nu t = A \sin \omega t. \tag{7.98}$$

Psi (ψ) is a very common symbol for wave amplitude. The *frequency* ν, defined by

$$\nu = \frac{1}{\tau} \tag{7.99}$$

gives the number of oscillations per second, conventionally expressed in *hertz* (Hz). An alternative measure of frequency is the number of *radians per second*, ω. Since one cycle corresponds to 2π radians,

$$\omega = 2\pi \nu. \tag{7.100}$$

The upper strip in Figure 7.3 shows the profile of the sound wave at a single instant of time. The pressure or density of the air also has a sinusoidal shape. At some given instant of time t the deviation of pressure from the undisturbed presssure p_0 might be represented by

$$\psi(x) = A \sin\left(\frac{2\pi x}{\lambda}\right),$$ (7.101)

where λ is the *wavelength* of the sound, the distance between successive pressure maxima. Sound consists of *longitudinal* waves, in which the wave amplitude varies in the direction parallel to the wave's motion. By contrast, electromagnetic waves, such as light, are *transverse*, with their electric and magnetic fields oscillating *perpendicular* to the direction of motion. The speed of light in vacuum, $c \approx 3 \times 10^8$ m/s. The speed of sound in air is much slower, typically around 350 m/s (1100 ft/s or 770 miles/hr— known as *Mach 1* for jet planes). As you know, thunder is the sound of lightning. You see the lightning essentially instantaneously but thunder takes about 5 s to travel 1 mile. You can calculate how far away a storm is by counting the number of seconds between the lightning and the thunder. A wave (light or sound) traveling at a speed c moves a distance of one wavelength λ in the time of one period τ. This implies the general relationship between frequency and wavelength

$$\frac{\lambda}{\tau} = \lambda\nu = c,$$ (7.102)

valid for all types of wave phenomena. A trumpet playing the same sustained note produces a much richer sound than a tuning fork, as shown in Figure 7.4. *Fourier analysis* of a musical tone shows a superposition of the fundamental frequency ν_0 augmented by *harmonics* or *overtones*, which are integer multiples of ν_0.

7.8   Dirac Deltafunction

Recall that the Kronecker delta δ_{nm}, defined in Eq. (7.78), pertains to the discrete variables n and m. A useful application enables the reduction of a summation to a single term:

$$\sum_{n=0}^{\infty} f_n \delta_{nm} = f_m.$$ (7.103)

The analog of the Kronecker delta for continuous variables is the *Dirac deltafunction* $\delta(x - x_0)$, which has the defining property

$$\int_{-\infty}^{\infty} f(x)\delta(x - x_0)\, dx = f(x_0),$$ (7.104)

which includes the normalization condition

$$\int_{-\infty}^{\infty} \delta(x - x_0)\, dx = 1.$$ (7.105)

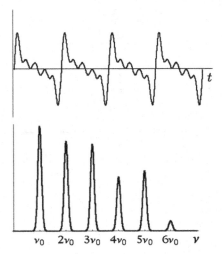

Figure 7.4 Fourier analysis of trumpet playing single note of frequency $\nu_0 \approx 523$ Hz, one octave above middle C. Upper curve represents the signal in the *time domain*, lower curve, in the *frequency domain*.

Evidently

$$\delta(x - x_0) \equiv \begin{cases} 0 & \text{if } x \neq x_0, \\ \infty & \text{if } x = x_0. \end{cases} \tag{7.106}$$

The approach to ∞ is sufficiently tame, however, that the integral has a finite value.

A simple representation for the deltafunction is the limit of a normalized Gaussian as the standard deviation approaches zero:

$$\delta(x - x_0) = \lim_{\sigma \to 0} \frac{1}{\sqrt{2\pi}\sigma} e^{-(x-x_0)^2/2\sigma^2}. \tag{7.107}$$

The Dirac deltafunction is shown pictorially in Figure 7.5. The deltafunction is the limit of a function which becomes larger and larger in an interval which becomes narrower and narrower. (Some university educators bemoaning increased specialization contend that graduate students are learning more and more about less and less until they eventually wind up knowing everything about nothing—the ultimate deltafunction!)

Differentiation of a function at a finite discontinuity produces a deltafunction. Consider, for example, the *Heaviside unit step function*:

$$H(x - x_0) \equiv \begin{cases} 0 & \text{if } x < x_0, \\ 1 & \text{if } x \geqslant x_0. \end{cases} \tag{7.108}$$

Sometimes $H(0)$ (for $x = x_0$) is defined as $\frac{1}{2}$. The derivative of the Heaviside function $H'(x - x_0)$ is clearly equal to zero when $x \neq x_0$. In addition

$$\int_{-\infty}^{\infty} H'(x - x_0)\, dx = H(\infty) - H(-\infty) = 1 - 0. \tag{7.109}$$

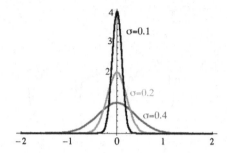

Figure 7.5 Dirac deltafunction with $x_0 = 0$ as limit of normalized Gaussian: $\delta(x) = \lim_{\sigma \to 0} \frac{1}{\sqrt{2\pi}\sigma} e^{-x^2/2\sigma^2}$

We can thus identify

$$H'(x - x_0) = \delta(x - x_0). \tag{7.110}$$

The deltafunction can be generalized to multiple dimensions. In three dimensions, the defining relation for a deltafunction can be expressed

$$\int f(\mathbf{r})\delta(\mathbf{r} - \mathbf{r}_0)d^3\mathbf{r} = f(\mathbf{r}_0). \tag{7.111}$$

For example, the limit of a continuous distribution of electrical charge $\rho(\mathbf{r})$ shrunken to a point charge q at $\mathbf{r}_0$ can be represented by

$$\rho(\mathbf{r}) = q\delta(\mathbf{r} - \mathbf{r}_0). \tag{7.112}$$

The potential energy of interaction between two continuous charge distributions is given by

$$V = -\int \int \frac{\rho_1(\mathbf{r}_1)\rho_2(\mathbf{r}_2)}{|\mathbf{r}_2 - \mathbf{r}_1|} d^3\mathbf{r}_1 d^3\mathbf{r}_2. \tag{7.113}$$

If the distribution $\rho_2(\mathbf{r}_2)$ is reduced to a point charge q_2 at $\mathbf{r}_2$, this reduces to

$$V = -q_2 \int \frac{\rho_1(\mathbf{r}_1)}{|\mathbf{r}_2 - \mathbf{r}_1|} d^3\mathbf{r}_1. \tag{7.114}$$

If the analogous thing then happens to $\rho_1(\mathbf{r}_1)$, the formula reduces to the Coulomb potential energy between two point charges

$$V = -\frac{q_1 q_2}{r_{12}}, \quad \text{where } r_{12} = |\mathbf{r}_2 - \mathbf{r}_1|. \tag{7.115}$$

Problem 7.8.1. Show that the normalized Lorentzian distribution

$$f(x) = \frac{1}{\pi} \frac{\gamma}{x^2 + \gamma^2}$$

approaches a deltafunction in the limit as $\gamma \to 0$.

7.9 Fourier Integrals

Fourier series are ideal for periodic functions, sums over frequencies which are integral multiples of some $2\pi/L$. For a more general class of functions which are not simply periodic, *all* possible frequency contributions must be considered. This can be accomplished by replacing a discrete Fourier series by a continuous integral. The coefficients c_m (or, equivalently, a_m and b_m) which represent the relative weight of each harmonic will turn into a *Fourier transform* $F(k)$, which measures the contribution of a frequency in a continuous range of k. In the limit as $L \to \infty$, a complex Fourier series (7.95) generalizes to a *Fourier integral*. The discrete variable $2\pi m/L$ can be replaced by a continuous variable k, such that

$$f(x) = \int_{-\infty}^{\infty} F(k)e^{ikx}\,dk \tag{7.116}$$

with the substitution

$$c_m \to \frac{2\pi}{L} F(k). \tag{7.117}$$

Correspondingly, Eq. (7.96) becomes

$$F(k) = \frac{1}{2\pi} \int_{-\infty}^{\infty} f(x)e^{-ikx}\,dx, \tag{7.118}$$

where $F(k)$ is called the Fourier transform of $f(x)$—alternatively written $g(k)$, $\tilde{f}(k)$, $\hat{f}(k)$, $\mathcal{F}[f]$ or sometimes simply $f(k)$. A Fourier-transform pair $f(x)$ and $F(k)$ can also be defined more symmetrically by writing:

$$f(x) = \frac{1}{\sqrt{2\pi}} \int_{-\infty}^{\infty} F(k)e^{ikx}\,dk, \quad F(k) = \frac{1}{\sqrt{2\pi}} \int_{-\infty}^{\infty} f(x)e^{ikx}\,dx. \tag{7.119}$$

Fourier integrals in the time-frequency domain have the form

$$f(t) = \int_{-\infty}^{\infty} F(\omega)e^{i\omega t}\,d\omega. \tag{7.120}$$

Figure 7.4 is best described as a Fourier transform of a trumpet tone since the spectrum of frequencies consists of peaks of finite width.

Another representation of the deltafunction, useful in the manipulation of Fourier transforms, is defined by the limit:

$$\delta(x) = \lim_{k \to \infty} \frac{\sin kx}{\pi x}. \tag{7.121}$$

For $x = 0$, the function equals k, which approaches ∞. For $x \neq 0$ the sine function oscillates with ever-increasing frequency as $k \to \infty$. The positive and negative contributions cancel so that the function becomes essentially equivalent to 0 under the integral sign. Finally, since

$$\int_{-\infty}^{\infty} \frac{\sin kx}{\pi x} = 1 \tag{7.122}$$

for finite values of k, Eq. (7.121) is suitably normalized to represent a deltafunction. The significance of this last representation is shown by the integral

$$\int_{-\infty}^{\infty} e^{i(k-k_0)x}\,dx = \lim_{X\to\infty}\int_{-X}^{X} e^{i(k-k_0)x}\,dx = \lim_{X\to\infty}\left[\frac{e^{i(k-k_0)X} - e^{-i(k-k_0)X}}{i(k-k_0)}\right]$$

$$= \lim_{X\to\infty}\left[\frac{2\sin(k-k_0)X}{(k-k_0)}\right] = 2\pi\delta(k-k_0). \tag{7.123}$$

This shows that the Fourier transform of a complex monochromatic wave $(2\pi)^{-1}e^{i(k-k_0)x}$ is a deltafunction $\delta(k-k_0)$.

An important result for Fourier transforms can be derived using (7.123). Using the symmetrical form for the Fourier integral (7.119), we can write

$$\int_{-\infty}^{\infty} |f(x)|^2\,dx = \int_{-\infty}^{\infty}\left[\frac{1}{\sqrt{2\pi}}\int_{-\infty}^{\infty} F(k)e^{ikx}\,dk\right.$$

$$\left. \times\frac{1}{\sqrt{2\pi}}\int_{-\infty}^{\infty} F^*(k')e^{-ik'x}\,dk'\right]dx \tag{7.124}$$

being careful to use the dummy variable k' in the second integral on the right. The integral over x on the right then gives

$$\int_{-\infty}^{\infty} e^{i(k-k')x}\,dx = 2\pi\delta(k-k'). \tag{7.125}$$

Following this by integration over k' we obtain

$$\int_{-\infty}^{\infty} |f(x)|^2\,dx = \int_{-\infty}^{\infty} |F(k)|^2\,dk, \tag{7.126}$$

a result known as *Parseval's theorem*. A closely related result is *Plancherel's theorem*:

$$\int_{-\infty}^{\infty} f(x)g^*(x)dx = \int_{-\infty}^{\infty} F(k)G^*(k)dk, \tag{7.127}$$

where F and G are the symmetric Fourier transforms of f and g, respectively.

The *convolution* of two functions is defined by

$$f * g \equiv \int_{-\infty}^{\infty} f(x')g(x-x')dx' = \int_{-\infty}^{\infty} f(x-x')g(x')dx'. \tag{7.128}$$

The Fourier transform of a convolution integral can be found by substituting the symmetric Fourier transforms F and G and using the deltafunction formula (7.125). The result is the *convolution theorem*, which can be expressed very compactly as

$$\mathcal{F}[f * g] = \mathcal{F}[f]\mathcal{F}[g]. \tag{7.129}$$

In an alternative form, $\mathcal{F}[fg] = \mathcal{F}[f] * \mathcal{F}[g]$.

7.10   Generalized Fourier Expansions

For Bessel functions and other types of special functions to be introduced later, it is possible to construct *orthonormal sets* of basis functions $\{\phi_n(x)\}$ which satisfy the orthogonality and normalization conditions:

$$\int \phi_n^*(x)\phi_{n'}(x)dx = \delta_{nn'} \tag{7.130}$$

with respect to integration over appropriate limits. (If the functions are real, complex conjugation is unnecessary.) An arbitrary function $f(x)$ in the same domain as the basis functions can be expanded in an expansion analogous to a Fourier series

$$f(x) = \sum_n c_n \phi_n(x) \tag{7.131}$$

with the coefficients determined by

$$c_n = \int \phi_n^*(x) f(x)dx. \tag{7.132}$$

If (7.132) is substituted into (7.131), with the appropriate use of a dummy variable, we obtain

$$f(x) = \sum_n \left[\int \phi_n^*(x') f(x')\, dx' \right] \phi_n(x) = \int \left[\sum_n \phi_n^*(x')\phi_n(x) \right] f(x')dx'. \tag{7.133}$$

The last quantity in square brackets has the same effect as the deltafunction $\delta(x - x')$. The relation known as *closure*

$$\sum_n \phi_n^*(x')\phi_n(x) = \delta(x - x') \tag{7.134}$$

is, in a sense, complementary to the orthonormality condition (7.130).

Generalized Fourier series find extensive application in mathematical physics, particularly quantum mechanics.

7.11   Asymptotic Series

In certain circumstances, a *divergent* series can be used to determine approximate values of a function as $x \to \infty$. Consider, as an example, the complementary error function erfc x, defined in Eq. (6.95):

$$\text{erfc } x = \frac{2}{\sqrt{\pi}} \int_x^\infty e^{-t^2}\, dt. \tag{7.135}$$

Noting that

$$e^{-t^2} \, dt = -\frac{1}{2t} \, d(e^{-t^2}). \tag{7.136}$$

Equation (7.135) can be integrated by parts to give

$$\text{erfc } x = \frac{2}{\sqrt{\pi}} \left(\frac{e^{-x^2}}{2x} - \int_x^\infty \frac{e^{-t^2}}{2t^2} \, dt \right) \tag{7.137}$$

Integrating by parts again:

$$\text{erfc } x = \frac{2}{\sqrt{\pi}} \left(\frac{e^{-x^2}}{2x} - \frac{e^{-x^2}}{4x^3} + \int_x^\infty \frac{3e^{-t^2}}{4t^4} \, dt \right). \tag{7.138}$$

Continuing the process, we obtain

$$\text{erfc } x \sim \frac{2e^{x^2}}{\sqrt{\pi}} \left[\frac{1}{2x} - \frac{1}{2^2 x^3} + \frac{1 \cdot 3}{2^3 x^5} - \frac{1 \cdot 3 \cdot 5}{2^4 x^7} + \cdots \right]. \tag{7.139}$$

This is an instance of an *asymptotic series*, as indicated by the equivalence symbol $\sim$ rather than equal sign. The series in brackets is actually divergent. However, a finite number of terms gives an approximation to erfc x for large values of x. The omitted terms, when expressed in their original form as an integral, as in Eqs. (7.137) or (7.138), approach zero as $x \to \infty$.

Consider the general case of an asymptotic series

$$f(x) \sim c_0 + \frac{c_1}{x} + \frac{c_2}{x^2} + \cdots = \sum_{k=0}^\infty \frac{c_k}{x^k} \equiv S(x) \tag{7.140}$$

with a partial sum

$$S_n(x) = \sum_{k=0}^n \frac{c_k}{x^k}. \tag{7.141}$$

The condition for $S(x)$ to be an asymptotic representation for $f(x)$ can be expressed

$$\lim_{x \to \infty} x^n [f(x) - S_n(x)] = 0 \quad \Rightarrow \quad f(x) \sim S(x). \tag{7.142}$$

A convergent series approaches $f(x)$ for a given x as $n \to \infty$, where n is the number of terms in the partial sum S_n. By contrast, an asymptotic series approaches $f(x)$ as $x \to \infty$ for a given n.

The *exponential integral* is another function defined as a definite integral which cannot be evaluated in a simple closed form. In the usual notation

$$-Ei(-x) \equiv \int_x^\infty \frac{e^{-t}}{t} \, dt. \tag{7.143}$$

By repeated integration by parts, the following asymptotic series for the exponential integral can be derived:

$$-Ei(-x) \sim \frac{e^{-x}}{x}\left(1 - \frac{1}{x} + \frac{2!}{x^2} - \frac{3!}{x^3} + \cdots\right). \tag{7.144}$$

Finally, we will derive an asymptotic expansion for the gamma function, applying a technique known as the *method of steepest descents*. Recall the integral definition:

$$x! = \Gamma(x+1) = \int_0^\infty t^x e^{-t}\, dt. \tag{7.145}$$

For large x the integrand $t^x e^{-t}$ will be very sharply peaked around $t = x$. It is convenient to write

$$t^x e^{-t} = e^{f(t)} \quad \text{with} \quad f(t) = x \ln t - t. \tag{7.146}$$

A Taylor series expansion of $f(t)$ about $t = x$ gives

$$f(t) = f(x) + (t-x)f'(x) + \frac{(t-x)^2}{2} f''(x) + \cdots$$

$$= x \ln x - x - \frac{(t-x)^2}{2x} + \cdots \tag{7.147}$$

noting that $f'(x) = 0$. The integral (7.145) can then be approximated as follows:

$$\Gamma(x+1) = \int_0^\infty e^{f(t)}\, dt \approx x^x e^{-x} \int_0^\infty e^{-(t-x)^2/2x}\, dt. \tag{7.148}$$

For large x, we introduce a negligible error by extending the lower integration limit to $-\infty$. The integral can then be done exactly to give *Stirling's formula*

$$= \Gamma(x+1) \approx \sqrt{2\pi x}\, x^x e^{-x} \quad \text{for } x \to \infty \tag{7.149}$$

or in terms of the factorial

$$n! \approx \sqrt{2\pi n}\, n^n e^{-n} = \sqrt{2\pi n}\left(\frac{n}{e}\right)^n \quad \text{for } n \to \infty. \tag{7.150}$$

This is consistent with the well-known approximation for the natural logarithm:

$$\ln(n!) \approx n \ln n - n. \tag{7.151}$$

A more complete asymptotic expansion for the gamma function will have the form

$$\Gamma(x+1) \sim \sqrt{2\pi}\, x^{x+\frac{1}{2}} e^{-x}\left(1 + \frac{c_1}{x} + \frac{c_2}{x^2} + \cdots\right). \tag{7.152}$$

Making use of the recursion relation

$$\Gamma(x+1) = x\Gamma(x) \tag{7.153}$$

it can be shown that $c_1 = 1/12$ and $c_2 = 1/288$.

You will notice that, as we go along, we are leaving more and more computational details for you to work out on your own. Hopefully, your mathematical facility is improving at a sufficient rate to keep everything understandable.

8 Differential Equations

To paraphrase the Greek philosopher Heraclitus,"The only thing constant is change." Differential equations describe mathematically how things in the physical world change—both in time and in space. While the solution of an algebraic equation is usually a *number*, the solution of a differential equation gives a *function*.

In this chapter, we consider *ordinary differential equations* (ODEs), which involve just two variables, say, an independent variable x and a dependent variable y. Later we will take up *partial differential equations*, which have two or more independent variables. The most general ordinary differential equation is a relationship of the form

$$F\left[x, y(x), \frac{dy}{dx}, \frac{d^2y}{dx^2}, \dots\right] = 0 \tag{8.1}$$

in which the object is to solve for $y(x)$, consistent with one or more specified *boundary conditions*. The *order* of a differential equation is determined by the highest-order derivative. For a large number of applications in physics, chemistry, and engineering, it suffices to consider first-and second-order differential equations, with no higher derivatives than $y''(x)$. In this chapter, we will deal with *linear* ODEs of the general form

$$y''(x) + p(x)y'(x) + q(x)y(x) = f(x), \tag{8.2}$$

where $p(x)$, $q(x)$, and $f(x)$ are known functions. When $f(x) = 0$, the equation is said to be *homogeneous*, otherwise *inhomogeneous*.

8.1 First-Order Differential Equations

A first-order equation of the form

$$\frac{dy}{dx} + q(x)y = 0 \tag{8.3}$$

can be solved by *separation of variables*. We rearrange the equation to

$$\frac{dy}{y} + q(x)dx = 0 \tag{8.4}$$

and then integrate to give

$$\int \frac{dy}{y} + \int q(x)dx = 0 \;\Rightarrow\; \ln y + \int q(x)dx = \text{const.} \tag{8.5}$$

Guide to Essential Math 2e. http://dx.doi.org/10.1016/B978-0-12-407163-6.00008-4

After exponentiating the equation (making it the exponent of e), we obtain

$$y(x) = \text{const} \times \exp\left(-\int q(x)dx\right).$$ (8.6)

The constant is determined by an initial or boundary condition.

For the inhomogeneous analog of (8.3):

$$\frac{dy}{dx} + q(x)y = f(x)$$ (8.7)

separation of variables can be done after introducing an *integrating factor*. Note first that

$$\frac{d}{dx}\left[y\exp\left(\int q(x)dx\right)\right] = \exp\left(\int q(x)dx\right)\left[\frac{dy}{dx} + q(x)y\right].$$ (8.8)

Thus Eq. (8.7) can be arranged to

$$\frac{d}{dx}\left[y\exp\left(\int q(x)dx\right)\right] = f(x)\exp\left(-\int q(x)dx\right).$$ (8.9)

The solution is

$$y(x) = e^{-\int q(x)dx}\left[f(x)e^{\int q(x)dx}\,dx + \text{const}\right].$$ (8.10)

This will look a lot less intimidating once the function $q(x)$ is explicitly specified.

We will consider next a few illustrative examples of first-order differential equations of physical significance. In several cases, independent variable x will be the time t.

A radioactive element disintegrates at a rate which is independent of any external conditions. The number of disintegrations occurring per unit time is proportional to the number of atoms originally present. This can be formulated as a first-order differential equation in time:

$$\frac{dN}{dt} = -kN(t).$$ (8.11)

The *rate constant* k is a characteristic of the radioactive element. The minus sign reflects the fact that the number of atoms N *decreases* with time. Equation (8.11) is easily solved by separation of variables. We find

$$\frac{dN}{N} = -k\,dt.$$ (8.12)

We now evaluate the indefinite integral of each side of the equation:

$$\int \frac{dN}{N} = -k\int dt \;\Rightarrow\; \ln N = -kt + \text{const}.$$ (8.13)

The constant of integration is, in this case, determined by the *initial condition*: at time $t = 0$, there are $N = N_0$ atoms. Exponentiating, we obtain

$$N(t) = N_0\,e^{-kt},$$ (8.14)

which is the famous equation for exponential decay. Radioactive elements are usually characterized by their *half-life* $t_{1/2}$, the time it takes for half the atoms to disintegrate. Setting $N(t_{1/2}) = \frac{1}{2}N_0$, we find the relation between half-life and decay constant:

$$t_{1/2} = \frac{\ln 2}{k}. \tag{8.15}$$

The result for radioactive decay is easily adapted to describe *organic growth*, the rate at which the population of a bacterial culture will increase given unlimited nutrition. We simply change $-k$ to $+k$ in Eq. (8.11), which leads to *exponential growth*:

$$N(t) = N_0 e^{kt}. \tag{8.16}$$

This remains valid until the source of nutrition runs out or the population density becomes intolerable.

A bimolecular chemical reaction such as $A + B \rightarrow C$ often follows a rate law

$$\frac{d[C]}{dt} = k[A][B] \tag{8.17}$$

in which the increase in concentration of the reaction product C is proportional to the product (in the sense of $\times$) of concentrations of the reactants A and B. To put this into more concrete form, assume that the initial concentrations at $t = 0$ are $[C]_0 = 0$, $[A]_0 = [B]_0 = a$. At a later time t, the concentration $[C]$ grows to x while $[A]$ and $[B]$ are reduced to $a - x$. Equation (8.17) then reduces to

$$\frac{dx}{dt} = k(a - x)^2. \tag{8.18}$$

Separation of variables followed by integration gives

$$\int \frac{dx}{(a - x)^2} = -\int \frac{d(a - x)}{(a - x)^2} = \frac{1}{a - x} = kt + \text{const.} \tag{8.19}$$

Since $x = 0$ when $t = 0$, the constant equals $1/a$ and the solution can be written

$$x(t) = \frac{a^2 kt}{1 + akt}. \tag{8.20}$$

More challenging is the case when $[B]_0 = b$, different from a. If you are adventurous, you can work out the solution

$$x(t) = \frac{ab[1 - e^{(a-b)kt}]}{b - ae^{(a-b)kt}}. \tag{8.21}$$

Problem 8.1.1. Find the solution to the differential equation

$$\frac{dy}{dx} = x^2 - 1,$$

satisfying the boundary condition $y(0) = 1$.

Problem 8.1.2. Solve the differential equation

$$\frac{dy}{dx} + y = e^x.$$

You will need to introduce an integrating factor.

Bernoulli's differential equation:

$$y' + q(x)y = f(x)y^n \tag{8.22}$$

is a well-known nonlinear equation that can be reduced to a linear equation. Multiplying by $(1-n)y^{-n}$, we obtain

$$(1-n)y^{-n}y' + (1-n)q(x)y^{1-n} = (1-n)f(x). \tag{8.23}$$

This reduces to a linear equation in the variable $z = y^{1-n}$:

$$z' + (1-n)q(x)z = (1-n)f(x). \tag{8.24}$$

Using the integrating factor $\exp[(1-n)\int q(x)dx]$, the general solution, after reverting to the original variable y, is given by

$$y^{1-n}e^{(1-n)\int q(x)dx} = (1-n)\int f(x)e^{(1-n)\int q(x)dx}\,dx + \text{const.} \tag{8.25}$$

The equation is already linear if $n = 0$ and separable, as well, if $n = 1$.

8.2 Numerical Solutions

Numerical methods can be used to obtain approximate solutions for differential equations that cannot be treated analytically. The most rudimentary procedure for first-order equations is the *Euler method*. Not, in itself, very accurate it serves as a starting point for more advanced techniques. Suppose we seek a solution of a first-order equation of the form $y'(x) = f(x, y)$, subject to the boundary condition $y(x_0) = y_0$. The first approximation for the value of y' at the point (x_1, y_1) might be suggested by the starting point in the definition of a derivative

$$y'(x_1) = \left.\frac{dy}{dx}\right|_1 \approx \frac{f(x_1, y_1) - f(x_0, y_0)}{x_1 - x_0}. \tag{8.26}$$

Choose a value for the size of each step between successive x-values, $h = \Delta x = x_1 - x_0$, so that $x_n = x_0 + nh$. Thus $y_1 = y_0 + hf(x_0, y_0)$. Each successive step in the Euler method is then determined by the recursive relations:

$$y_{n+1} = y_n + hf(x_n, y_n). \tag{8.27}$$

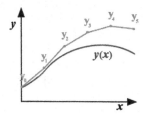

Figure 8.1 Pictorial representation of the Euler method.

This procedure is illustrated by the chain of red[1] segments in Figure 8.1, with the blue curve representing the hypothetical exact solution. Clearly, the accuracy degrades with increasing x, although it can be improved by decreasing the interval size h.

Runge-Kutta methods provide more advanced approaches to numerical integration of ordinary differential equations. The most common variant, known as RK4, is an elaboration of the Euler method using four-term interpolations to determine successive points on the solution curve. Intermediate values of the functions $f(x, y)$ between x_n and x_{n+1} enter into the computation of successive coefficients. The RK4 method is based on the following equations:

$$y_{n+1} = y_n + \frac{1}{6}(k_1 + 2k_2 + 2k_3 + k_4) \tag{8.28}$$

with

$$k_1 = hf(x_n, y_n), \tag{8.29}$$

$$k_2 = hf(x_n + \frac{1}{2}h, y_n + \frac{1}{2}k_1), \tag{8.30}$$

$$k_3 = hf(x_n + \frac{1}{2}h, y_n + \frac{1}{2}k_2), \tag{8.31}$$

$$k_4 = hf(x_n + h, y_n + k_3). \tag{8.32}$$

When $f(x, y)$ is a function of x alone, the differential equation is essentially just an integration and the procedure reduces to Simpson's rule. The RK4 method has an accumulated error of order h^4, whereas the Euler method had an error which was first order in h.

Mathematica is particularly adept at numerical solutions of differential equations, automatically choosing the most appropriate algorithm. For example, a first-order equation of the type discussed above can be programed using a command something like:
```
NDSolve[{y[x]==..., y[a]==...},{x, a, b}].
```

8.3   AC Circuits

A number of instructive applications of differential equations concern alternating-current (AC) circuits containing resistance, inductance, capacitance and an oscillating

---

[1]For interpretation of color in Figure 8.1, the reader is referred to the web version of this book.

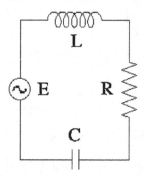

Figure 8.2 Series RLC circuit powered by an oscillating emf.

voltage source, as represented in Figure 8.2. In the simplest case, consider a circuit with resistance R and voltage (or emf) E. The current I is determined by *Ohm's law*:

$$I = \frac{E}{R}. \tag{8.33}$$

The standard units are amperes for I, volts for E, and ohms for R. The other relevant units are henrys for L and farads for C. Ohm's law is true even for an AC circuit, in which the voltage varies sinusoidally with time, say

$$E(t) = E_0 \cos \omega t. \tag{8.34}$$

The current is then given by

$$I(t) = \frac{E(t)}{R} = \frac{E_0}{R} \cos \omega t = I_0 \cos \omega t. \tag{8.35}$$

Thus the current through a resistance oscillates *in phase* with the voltage. The frequency ω is expressed in units of rad/s. It is more common to measure frequency ν in *cycles/s*, a unit called the hertz (Hz). Since one cycle traces out 2π radians, the two measures of frequency are related by

$$\omega = 2\pi \nu. \tag{8.36}$$

Thus your 60 Hz household voltage has an angular frequency of $\omega = 2\pi \times 60 \approx 377$ rad/s.

The voltage change across an inductance is given by $L\, dI/dt$. Thus for a circuit with inductance, but negligible resistance, the analog of Ohm's law is

$$L\frac{dI}{dt} = E. \tag{8.37}$$

With an oscillating voltage (8.34), with an initial current $I(0) = 0$, this equation is easily integrated to give

$$I(t) = \frac{E_0}{\omega L} \sin \omega t = \frac{E_0}{X_L} \cos(\omega t - \pi/2), \tag{8.38}$$

where the *inductive reactance* $X_L = \omega L$ has the same units as R. For a DC voltage ($\omega = 0$), an inductor behaves just like an ordinary conductor. Note that the current in Eq. (8.38) is 90° out of phase with the voltage $E(t) = E_0 \cos \omega t$. Specifically, for a pure inductance, the current *lags* the voltage by $\pi/2$. Alternatively stated, the voltage *leads* the current by $\pi/2$. Physically, this reflects the fact that the inductor builds up an opposing emf (by Lenz's law) in response to an increase in current.

For a circuit with capacitance C, the relevant relation is

$$E = \frac{q}{C}, \tag{8.39}$$

where q (in coulombs) is the charge on the capacitor. In a DC circuit, no current passes through a capacitor. For an AC circuit, however, with the current being given by $I = dq/dt$, we find

$$\frac{dE}{dt} = \frac{I(t)}{C}. \tag{8.40}$$

For an AC voltage (8.34), the equation integrates to give

$$I(t) = -\omega C E_0 \sin \omega t = \frac{E_0}{X_C} \cos(\omega t + \pi/2), \tag{8.41}$$

where the *capacitive reactance* is defined by $X_C = 1/\omega C$. This shows that for a pure capacitance, the current *leads* the voltage by $\pi/2$. The mnemonic "ELI the ICEman" summarizes the phase relationships for inductance and capacitance: for L, E leads I while for C, I leads E.

In an electrical circuit with both resistance and inductance, the current and voltage are related by

$$L\frac{dI}{dt} + RI = E. \tag{8.42}$$

Suppose at time $t = 0$, while the current has the value I_0, the voltage E is suddenly turned off. With $E = 0$, Eq. (8.42) reduces to the form of Eq. (8.11). Thus the current must decay exponentially with

$$I(t) = I_0 e^{-kt} \quad \text{where } k \equiv \frac{R}{L}. \tag{8.43}$$

As a more challenging problem, let the circuit be powered by an AC voltage $E(t) = E_0 \cos \omega t$. Now (8.42) becomes an inhomogeneous equation.

A very useful trick when dealing with quantities having sinusoidal dependence takes advantage of Euler's theorem (4.57) in the form

$$e^{i\omega t} = \cos \omega t + i \sin \omega t. \tag{8.44}$$

Note that

$$\cos \omega t = \Re\, e^{i\omega t} \quad \text{and} \quad \sin \omega t = \Im\, e^{i\omega t}. \tag{8.45}$$

Exponentials are much simpler to differentiate and integrate than sines and cosines. Suppose the AC voltage above is imagined to have a complex form $\mathcal{E}(t) = E_0\, e^{i\omega t}$. At the end of the computation, we take the real part of the resulting equation. The answer will be the same, as if we had used $E(t) = E_0 \cos \omega t$ throughout. (A word of caution: make certain that the complexified functions occur only *linearly* in the equations.)

Accordingly, we write the circuit equation

$$\frac{d\mathcal{I}}{dt} + k\mathcal{I}(t) = \frac{E_0}{L} e^{i\omega t}, \tag{8.46}$$

where $\mathcal{I}$ is a complex variable representing the current. We can separate variables in this equation by introducing the integrating factor e^{kt}. Since

$$\frac{d}{dt}\left(\mathcal{I}e^{kt}\right) = e^{kt}\left(\frac{d\mathcal{I}}{dt} + k\mathcal{I}\right). \tag{8.47}$$

Equation (8.46) reduces to

$$\frac{d}{dt}\left(\mathcal{I}e^{kt}\right) = \frac{E_0}{L} e^{kt}\, e^{i\omega t}. \tag{8.48}$$

Integration gives

$$\mathcal{I}e^{kt} = \frac{E_0}{L}\int e^{(k+i\omega)t}\, dt = \frac{E_0}{L}\frac{e^{(k+i\omega)t}}{(k+i\omega)} + \text{const.} \tag{8.49}$$

If we specify that $I = 0$ when $t = 0$, we find

$$\mathcal{I}(t) = \frac{E_0}{L}\frac{e^{i\omega t}}{(k+i\omega)} - \frac{E_0}{L}\frac{e^{-kt}}{(k+i\omega)}. \tag{8.50}$$

The physically significant result is the real part:

$$I(t) = \Re\mathcal{I}(t) = \frac{1}{2}[\mathcal{I}(t) + \mathcal{I}^*(t)] = \frac{E_0}{L}\frac{k\cos\omega t + \omega \sin\omega t}{k^2 + \omega^2} - \frac{E_0}{L}\frac{ke^{-kt}}{k^2 + \omega^2}. \tag{8.51}$$

The last term represents a transient current, which damps out for $kt \gg 1$. The terms which persist represent the *steady-state solution*. Note that in the DC limit, as

$\omega \to 0$, the solution reduces to Ohm's law $I = E/R$. The steady-state solution can be rearranged to the form

$$I(t) = \frac{E_0}{Z} \cos(\omega t - \delta).$$ (8.52)

Here the *impedance* of the RL circuit is defined by $Z = \sqrt{R^2 + X_L^2}$ while the phase shift δ is given by $\tan \delta = \omega/k = X_L/R$.

The steady-state solution can be deduced directly from Eq. (8.46) by assuming from the outset that $\mathcal{I}(t) = \mathcal{I}_0 e^{i\omega t}$. The equation can then be readily solved to give

$$\mathcal{I}_0 = \frac{E_0}{L} \frac{1}{(k + i\omega)},$$ (8.53)

which is equivalent to (8.50). One can also define a complex impedance $\mathcal{Z} = R + iX_L = Z e^{i\delta}$, in terms of which we can write a complex generalization of Ohm's law:

$$\mathcal{I} = \frac{\mathcal{E}}{\mathcal{Z}}.$$ (8.54)

Circuits with R, L, and C involve second-order differential equations, which is our next topic.

8.4 Second-Order Differential Equations

We will consider here linear second-order equations with constant coefficients, in which the functions $p(x)$ and $q(x)$ in Eq. (8.2) are constants. The more general case gives rise to *special functions*, several of which we will encounter later as solutions of partial differential equations. The homogeneous equation, with $f(x) = 0$, can be written

$$\frac{d^2 y}{dx^2} + a_1 \frac{dy}{dx} + a_2 y = 0.$$ (8.55)

It is convenient to define the *differential operator*

$$D \equiv \frac{d}{dx}$$ (8.56)

in terms of which

$$Dy = \frac{dy}{dx} \quad \text{and} \quad D^2 y = \frac{d^2 y}{dx^2}.$$ (8.57)

The differential equation (8.55) is then written

$$D^2 y + a_1 Dy + a_2 y = 0,$$ (8.58)

or, in factored form,

$$(D - r_1)(D - r_2)y = (D - r_2)(D - r_1)y = 0,$$ (8.59)

where r_1, r_2 are the roots of the *auxilliary equation*

$$r^2 + a_1 r + a_2 = 0. \tag{8.60}$$

The solutions of the two first-order equations

$$(D - r_1)y = 0 \quad \text{or} \quad \frac{dy}{dx} + r_1 y = 0 \tag{8.61}$$

give

$$y = \text{const } e^{r_1 x}, \tag{8.62}$$

while

$$(D - r_2)y = 0 \quad \text{or} \quad \frac{dy}{dx} + r_2 y = 0 \tag{8.63}$$

gives

$$y = \text{const } e^{r_2 x}. \tag{8.64}$$

Clearly, these are also solutions to Eq. (8.59). The general solution is the linear combination

$$y(x) = c_1 e^{r_1 x} + c_2 e^{r_2 x}. \tag{8.65}$$

In the case that $r_1 = r_2 = r$, one solution is apparently lost. We can recover a second solution by considering the limit:

$$\lim_{r_1 \to r_2} \frac{e^{r_2 x} - e^{r_1 x}}{r_2 - r_1} = \frac{\partial}{\partial r} e^{r x} = x e^{r x}. \tag{8.66}$$

(Remember that the partial derivative $\partial/\partial r$ does the same thing as d/dr, with every other variable held constant.) Thus the general solution for this case becomes

$$y(x) = c_1 e^{r x} + c_2 x e^{r x}. \tag{8.67}$$

When r_1 and r_2 are imaginary numbers, say ik and $-ik$, the solution (8.65) contains complex exponentials. Since, by Euler's theorem, these can be expressed as sums and difference of sine and cosine, we can write

$$y(x) = c_1 \cos kx + c_2 \sin kx. \tag{8.68}$$

Many applications in physics, chemistry, and engineering involve a simple differential equation, either

$$y''(x) + k^2 y(x) = 0 \quad \text{or} \quad y''(x) - k^2 y(x) = 0. \tag{8.69}$$

The first equation has trigonometric solutions $\cos kx$ and $\sin kx$, while the second has exponential solutions e^{kx} and e^{-kx}. These results can be easily verified by "reverse engineering." For example, assuming that $y(x) = \cos kx$, then $y'(x) = -k \sin kx$ and $y''(x) = -k^2 \cos kx$. It follows that $y''(x) + k^2 y(x) = 0$.

8.5   Some Examples from Physics

Newton's second law of motion in one dimension has the form

$$f(x) = ma(t) = m\frac{dv}{dt} = m\frac{d^2x}{dt^2},$$ (8.70)

where f is the force on a particle of mass m, a is the acceleration, and v is the velocity. Newton's law leads to a second-order differential equation, with the solution $x(t)$ determining the motion of the particle. The simplest case is a free particle, with $f = 0$. Newton's law then reduces to

$$m\frac{d^2x}{dt^2} = 0.$$ (8.71)

The solution is

$$x(t) = x_0 + v_0 t$$ (8.72)

in which x_0 and v_0 are the two constants of integration, representing, respectively, the initial $(t = 0)$ position and velocity. Uniform linear motion at constant velocity v_0 is also in accord with Newton's *first* law of motion: a body in motion tends to remain in motion at constant velocity, unless acted upon by some external force.

A slightly more general problem is motion under a constant force. An example is a body in free fall in the general vicinity of the Earth's surface, which is subject to a constant downward force $f = -mg$. Here $g \approx 9.8$ m/s^2 (about 32 ft/s^2), the *acceleration of gravity*. Denoting the altitude above the Earth's surface by z, we obtain the differential equation

$$m\frac{d^2z}{dt^2} = -mg.$$ (8.73)

The factors m cancel. (This is actually a very profound result called the *equivalence principle*: the gravitational mass of a body equals its inertial mass. It is the starting point for Einstein's General Theory of Relativity.) One integration of Eq. (8.73) gives

$$\frac{dz}{dt} = -gt + v_0,$$ (8.74)

where the constant of integration v_0 represents the initial velocity. A second integration gives

$$z(t) = -\frac{1}{2}gt^2 + v_0 t + z_0,$$ (8.75)

where the second constant of integration z_0 represents the initial altitude. This solution is consistent with the well-known result that a falling body (neglecting air resistance) goes about 16 ft in 1 s, 64 ft after 2 s, and so on.

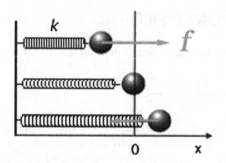

Figure 8.3 Hooke's law $f = -kx$ for a spring.

The force exerted by a metal spring subject to moderate stretching or compression is given approximately by *Hooke's law*

$$f = -kx. \tag{8.76}$$

Here x represents the displacement, positive or negative, from the spring's equilibrium extension, as shown in Figure 8.3. The *force constant k* is a measure of the spring's stiffness. The minus sign reflects the fact that the force is in the opposite direction to the displacement—a spring will resist equally either stretching and compression. Consider now a mass m connected to a spring with force constant k. Assume for simplicity that the mass of the spring itself is negligible compared to m. Newton's second law for this idealized system leads to the differential equation

$$m\frac{d^2x}{dt^2} + kx = 0. \tag{8.77}$$

The auxilliary equation (8.60) is

$$r^2 + \frac{k}{m} = 0 \tag{8.78}$$

with roots

$$r = \pm i\sqrt{\frac{k}{m}} = \pm i\omega_0 t, \qquad \omega_0 \equiv \sqrt{\frac{k}{m}}. \tag{8.79}$$

Thus the solution is a linear combination of complex exponentials

$$x(t) = c_1' e^{i\omega_0 t} + c_2' e^{-i\omega_0 t}. \tag{8.80}$$

Alternatively,

$$x(t) = c_1 \cos \omega_0 t + c_2 \sin \omega_0 t. \tag{8.81}$$

This shows that the spring oscillates sinusoidally with a *natural frequency* $\omega_0 = \sqrt{k/m}$. Sinusoidal oscillation is also called *harmonic motion* and the idealized system is referred to as a *harmonic oscillator*. If necessary, the two constants of integration can be determined by the initial displacement and velocity of the oscillating mass.

Another form for the solution (8.81) can be obtained by setting $c_1 = A \sin \delta$, $c_2 = A \cos \delta$. We find then

$$x(t) = A \sin(\omega_0 t + \delta). \tag{8.82}$$

An alternative possibility is $x(t) = A' \cos(\omega_0 t + \delta')$.

The oscillation of a real spring will eventually be damped out, in the absence of external driving forces. A reasonable approximation for damping is a force retarding the motion which is proportional to the instantaneous velocity: $f = -bv = -b \, dx/dt$, where b is called the *damping constant*. The differential equation for a damped harmonic oscillator can be written

$$\frac{d^2x}{dt^2} + 2\gamma \frac{dx}{dt} + \omega_0^2 x = 0, \tag{8.83}$$

where $\gamma \equiv b/2m$. The auxilliary equation has the roots $r = -\gamma \pm \sqrt{\gamma^2 - \omega_0^2}$ so that the general solution can be written:

$$x(t) = c_1 \exp\left[-\gamma t + (\omega_0^2 - \gamma^2)^{1/2} t\right] + c_2 \exp\left[-\gamma t - (\omega_0^2 - \gamma^2)^{1/2} t\right]. \tag{8.84}$$

For cases in which damping is not too extreme, so that $\omega_0 > \gamma$, the square roots are imaginary. The solution (8.84) can be written in the form

$$x(t) = A e^{-\gamma t} \cos\left[(\omega_0^2 - \gamma^2)^{1/2} t + \delta\right]. \tag{8.85}$$

This represents a damped sinusoidal wave, as shown in Figure 8.4. For stronger damping, such that $\gamma \geqslant \omega_0$, $x(t)$ decreases exponentially with no oscillation.

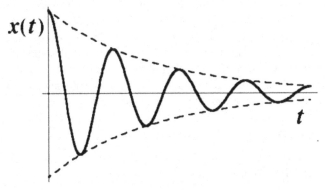

Figure 8.4 Damped sinusoidal wave of form $e^{-\gamma t} \cos[(\omega_0^2 - \gamma^2)^{1/2} t]$. The dotted lines represent the envelopes $\pm e^{-\gamma t}$.

A damped Hooke's-law system subject to *forced oscillations* at frequency ω can be represented by an inhomogeneous differential equation:

$$\frac{d^2x}{dt^2} + 2\gamma\frac{dx}{dt} + \omega_0^2 x = f_0 \cos \omega t. \tag{8.86}$$

Again it is useful to assume a complex exponential driving force $f_0 e^{i\omega t}$. If we seek, as before, just the steady-state solutions to (8.86), the complex form $\chi = \chi_0 e^{i\omega t}$ can be substituted for x. This reduces (8.86) to an algebraic equation for χ_0:

$$-\omega^2 \chi_0 + 2i\gamma\omega\chi_0 + \omega_0^2 \chi_0 = f_0 \tag{8.87}$$

with

$$\chi_0 = \frac{f_0}{\omega_0^2 - \omega^2 + 2i\gamma\omega}. \tag{8.88}$$

The steady-state solution is then given by

$$x(t) = \Re\left(\chi_0 e^{i\omega t}\right) = f_0 \frac{(\omega_0^2 - \omega^2)\cos\omega t - 2\gamma\omega\sin\omega t}{(\omega_0^2 - \omega^2)^2 + 4\gamma^2\omega^2}, \tag{8.89}$$

or, more compactly,

$$x(t) = \frac{f_0}{[(\omega_0^2 - \omega^2)^2 + 4\gamma^2\omega^2]^{1/2}} \cos(\omega t + \delta), \tag{8.90}$$

where

$$\tan\delta = \frac{2\gamma\omega}{\omega_0^2 - \omega^2}. \tag{8.91}$$

It can be seen that, after transients die out, the forced oscillation impresses its frequency ω on the system, apart from a phase shift δ. If this frequency matches the natural frequency of the oscillator, ω_0, the system is said to be in *resonance*. The amplitude of the oscillation then reaches a maximum. At resonance, $\delta = \pi/2$, so that the oscillator's motion is 90° out of phase with the forcing function.

The behavior of an RLC circuit (Figure 8.2) is closely analogous to that of an oscillating spring. The circuit equation (8.42) generalizes to

$$L\frac{dI}{dt} + RI + \frac{q}{C} = E. \tag{8.92}$$

Taking the time derivative leads to a more useful form:

$$L\frac{d^2I}{dt^2} + R\frac{dI}{dt} + \frac{I}{C} = \frac{dE}{dt}, \tag{8.93}$$

recalling that $I = dq/dt$. To obtain the steady-state solution, we assume the complex forms $\mathcal{E} = E_0\,e^{i\omega t}$ and $\mathcal{I} = \mathcal{I}_0\,e^{i\omega t}$. Equation (8.93) then reduces to an algebraic equation

$$\left(-\omega^2 L + i\omega R + \frac{1}{C}\right)\mathcal{I}_0 = i\omega E_0. \tag{8.94}$$

The result is most compactly expressed in terms of the impedance:

$$Z = \sqrt{R^2 + X^2} = \sqrt{R^2 + \left(\omega L - \frac{1}{\omega C}\right)^2}. \tag{8.95}$$

Here X is the *reactance* of the circuit, equal to the difference between inductive and capacitive reactance:

$$X = X_L - X_C = \omega L - \frac{1}{\omega C}. \tag{8.96}$$

In terms of the complex impedance

$$\mathcal{Z} = R + i\left(\omega L - \frac{1}{\omega C}\right) = Z\,e^{i\delta}, \quad \tan\delta = \frac{X}{R}. \tag{8.97}$$

Equation (8.94) can be solved for

$$\mathcal{I}_0 = \frac{E_0}{\mathcal{Z}}. \tag{8.98}$$

Therefore the physical value of the current is given by

$$I(t) = \Re(\mathcal{I}_0\,e^{i\omega t}) = \frac{E_0}{Z}\cos(\omega t - \delta). \tag{8.99}$$

The resonance frequency for an RLC circuit is determined by the condition

$$\omega L - \frac{1}{\omega C} = 0 \;\Rightarrow\; \omega = \frac{1}{\sqrt{LC}}. \tag{8.100}$$

At resonance, $X = 0$, $Z = R$, and $\delta = 0$. The inductive and capacitive reactances exactly cancel so that the impedance reduces to a pure resistance. Thus the current is maximized and oscillates in phase with the voltage. In a circuit designed to detect electromagnetic waves (e.g. radio or television signals) of a given frequency, the inductance and capacitance are "tuned" to satisfy the appropriate resonance condition.

8.6   Boundary Conditions

Often, the boundary conditions imposed on a differential equation determine significant aspects of its solutions. We consider two examples involving the one-dimensional Schrödinger equation in quantum mechanics:

$$-\frac{\hbar^2}{2m}\frac{d^2\psi}{dx^2} + V(x)\psi(x) = E\psi(x). \tag{8.101}$$

It will turn out that boundary conditions determine the values of E, the allowed energy levels for a quantum system. A *particle-in-a-box* is a hypothetical system with a potential energy given by

$$V(x) = \begin{cases} \infty & \text{for } x < 0 \text{ or } x > a, \\ 0 & \text{for } 0 \leqslant x \leqslant a. \end{cases} \tag{8.102}$$

Because of the infinite potential nergy, the wavefunction $\psi(x) = 0$ outside the box, where $x \leqslant 0$ or $x \geqslant a$. This provides two boundary conditions

$$\psi(0) = 0 \quad \text{and} \quad \psi(a) = 0 \tag{8.103}$$

for the Schrödinger eqution inside the box, which can be written in the familiar form

$$\psi''(x) + k^2 \psi(x) = 0, \tag{8.104}$$

where

$$k^2 \equiv \frac{2mE}{\hbar^2}. \tag{8.105}$$

The general solution to (8.104) can be written

$$\psi(x) = A \sin kx + B \cos kx. \tag{8.106}$$

Imposing the boundary condition at $x = 0$

$$\psi(0) = A \sin 0 + B \cos 0 = B = 0, \tag{8.107}$$

which reduces the general solution to

$$\psi(x) = A \sin kx. \tag{8.108}$$

The boundary condition at $x = a$ implies

$$\psi(a) = A \sin ka = 0. \tag{8.109}$$

We can't just set $A = 0$ because that would imply $\psi(x) = 0$ everywhere. Recall, however, that the sine function periodically goes through 0, when its argument equals $\pi, 2\pi, 3\pi, \ldots$ Therefore the second boundary condition can satisfied if

$$ka = n\pi, \qquad n = 1, 2, 3, \ldots \tag{8.110}$$

This implies that $k = n\pi/a$ and, by (8.105), the allowed values of the energy are then given by

$$E_n = \frac{\hbar^2}{2m} \frac{n^2 \pi^2}{a^2}, \qquad n = 1, 2, 3, \ldots, \tag{8.111}$$

where the integer n is called a *quantum number*. Quantization of energy in a bound quantum system is thus shown to be a consequence of boundary conditions imposed on the Schrödinger equation. The wavefunction corresponding to the energy E_n is given by

$$\psi(x) = A \sin\left(\frac{n\pi x}{a}\right). \tag{8.112}$$

Setting $A = (2/a)^{1/2}$, we obtain wavefunctions fulfilling the *normalization condition*:

$$\int_0^a [\psi_n(x)]^2 dx = 1. \tag{8.113}$$

Problem 8.6.1. Redo the analysis of the same problem with the alternative boundary conditions:

$$V(x) = \begin{cases} \infty & \text{for } x < -a \text{ or } x > a, \\ 0 & \text{for } -a \leqslant x \leqslant a. \end{cases}$$

For a *free particle* in quantum mechanics, $V(x) = 0$ everywhere. The Schrödinger equation (8.104) still applies, but now with no restrictive boundary conditions. Any value of k is allowed $(-\infty, k < \infty)$ and thus $E \geqslant 0$. There is no quantization of energy for a free particle. The wavefunction is conventionally written

$$\psi(x) = \text{const } e^{ikx}, \tag{8.114}$$

with $k > 0$ $[k < 0]$ corresponding to a particle moving to the right [left]. It is convenient, for bookkeeping purposes to impose *periodic boundary conditions*, such that

$$\psi(x + nL) = \psi(x), \qquad n = 0, \pm 1, \pm 2, \ldots \tag{8.115}$$

This requires that

$$e^{ik(x+L)} = e^{ikx} \Rightarrow e^{ikL} = 1, \tag{8.116}$$

which is satisfied if

$$kL = 2n\pi, \qquad n = 0, \pm 1, \pm 2, \ldots \tag{8.117}$$

This implies an artificial quantization of k with

$$k_n = \frac{2n\pi}{L}. \tag{8.118}$$

But, since L can be arbitrarily chosen, all values of k are allowed. With the constant in Eq. (8.114) set equal to L^{-1}, the wavefunction obeys the *box normalization* condition

$$\int_0^L \psi_n^*(x)\psi_n(x) dx = 1. \tag{8.119}$$

More generally, since the functions $\psi_n(x)$ and $\psi_n'(x)$ are orthogonal for $n \neq n'$, we can write

$$\int_0^L \psi_n^*(x)\psi_n'(x) dx = \delta_{nn'} \tag{8.120}$$

in terms of the Kronecker delta (7.78). The functions $\{L^{-1} e^{2n\pi x/L}, -\infty \leqslant n \leqslant \infty\}$ thus constitute an orthonormal set.

8.7   Series Solutions

A very general method for obtaining solutions to second-order differential equations is to expand $y(x)$ in a power series and then evaluate the coefficients term by term. We will illustrate the method with a trivial example which we have already solved, namely the equation with constant coefficients:

$$y''(x) + k^2 y(x) = 0. \tag{8.121}$$

Assume that $y(x)$ can be expanded in a power series about $x = 0$:

$$y(x) = \sum_{n=0}^{\infty} a_n x^n. \tag{8.122}$$

The first derivative is given by

$$y'(x) = \sum_{n=1}^{\infty} n a_n x^{n-1} \xrightarrow{n \to n+1} \sum_{n=0}^{\infty} (n+1) a_{n+1} x^n. \tag{8.123}$$

We have redefined the summation index in order to retain the dependence on x^n. Analogously,

$$y''(x) = \sum_{n=2}^{\infty} n(n-1) a_n x^{n-2} \xrightarrow{n \to n+2} \sum_{n=0}^{\infty} (n+2)(n+1) a_{n+2} x^n. \tag{8.124}$$

Equation (8.121) then implies

$$\sum_{n=0}^{\infty} \left[(n+2)(n+1) a_{n+2} + k^2 a_n \right] x^n = 0. \tag{8.125}$$

Since this is true for all values of x, every quantity in square brackets must equal zero. This leads to the *recursion relation*

$$a_{n+2} = -\frac{k^2}{(n+2)(n+1)} a_n. \tag{8.126}$$

Let a_0 be treated as one constant of integration. We then find

$$a_2 = -\frac{k^2}{2 \cdot 1} a_0, \qquad a_4 = -\frac{k^2}{4 \cdot 3}, \qquad a_2 = +\frac{k^4}{4!} a_0, \ \dots \tag{8.127}$$

It is convenient to rewrite the coefficient a_1 as $k a_1$. We find thereby

$$a_3 = -\frac{k^3}{3 \cdot 2} a_1, \qquad a_5 = +\frac{k^5}{5!} a_1, \qquad \dots \tag{8.128}$$

The general power-series solution of the differential equation is thus given by

$$y(x) = \left(1 - \frac{k^2x^2}{2!} + \frac{k^4x^4}{4!} - \cdots\right)a_0$$
$$+ \left(x - \frac{k^3x^3}{3!} + \frac{k^5x^5}{5!} - \cdots\right)a_1, \tag{8.129}$$

which is recognized as the expansion for

$$y(x) = a_0 \cos kx + a_1 \sin kx. \tag{8.130}$$

We consider next the more general case of a linear homogeneous second-order differential equation with nonconstant coefficients:

$$y''(x) + p(x)y'(x) + q(x)y(x) = 0. \tag{8.131}$$

If the functions $p(x)$ and $q(x)$ are both finite at $x = x_0$, then x_0 is called a *regular point* of the differential equation. If either $p(x)$ or $q(x)$ diverges as $x \to x_0$, then x_0 is called a *singular point*. If both $(x - x_0)p(x)$ and $(x - x_0)^2q(x)$ have finite limits as $x \to x_0$, then x_0 is called a *regular singular point* or *nonessential singularity*. If either of these limits continues to diverge, the point is an *essential singularity*.

For regular singular points, a series solution of the differential equation can be obtained by the *method of Frobenius*. This is based on the following generalization of the power-series expansion:

$$y(x) = x^\alpha \sum_{k=0}^{\infty} a_k x^k = \sum_{k=0}^{\infty} a_k x^{k+\alpha}. \tag{8.132}$$

The derivatives are then given by

$$y'(x) = \sum_{k=0}^{\infty}(\alpha + k)a_k x^{k+\alpha-1} \tag{8.133}$$

and

$$y''(x) = \sum_{k=0}^{\infty}(\alpha + k)(\alpha + k - 1)a_k x^{k+\alpha-2}. \tag{8.134}$$

The possible values of α are obtained from the *indicial equation*, which is based on the presumption that a_0 is the first nonzero coefficient in the series (8.132).

8.8 Bessel Functions

In later work we will encounter *Bessel's differential equation*:

$$x^2y''(x) + xy'(x) + (x^2 - n^2)y(x) = 0, \tag{8.135}$$

one of the classics of mathematical physics. Bessel's equation occurs, in particular, in a number of applications involving cylindrical coordinates. Dividing the standard form (8.135) by x^2 shows that $x = 0$ is a regular singular point of Bessel's equation. The method of Frobenius is thus applicable. Substituting the power-series expansion (8.132) into (8.135), we obtain

$$\sum_{k=0}^{\infty} \left[(\alpha + k)(\alpha + k - 1)a_k + (\alpha + k)a_k + a_{k-2} - n^2 a_k \right] x^{k+\alpha} = 0. \qquad (8.136)$$

This leads to the recursion relation

$$a_{k-2} = -[(\alpha + k)(\alpha + k - 1) + (\alpha + k) - n^2]a_k. \qquad (8.137)$$

Setting $k = 0$ in the recursion relation and noting that $a_{-2} = 0$ (a_0 is the first nonvanishing coefficient), we obtain the indicial equation

$$\alpha(\alpha - 1) + \alpha - n^2 = 0. \qquad (8.138)$$

The roots are $\alpha = \pm n$. With the choice $\alpha = n$, the recursion relation simplifies to

$$a_k = -\frac{a_{k-2}}{k(2n + k)}. \qquad (8.139)$$

Since $a_{-1} = 0$, $a_1 = 0$ (assuming $n \neq -\frac{1}{2}$). Likewise $a_3, a_5, \ldots = 0$, as do all odd a_k. For even k, we have

$$a_2 = -\frac{a_0}{2(2n + 2)}, \quad a_4 = -\frac{a_2}{4(2 + 4n)} = \frac{a_0}{2 \cdot 4(2n + 2)(2n + 4)}. \qquad (8.140)$$

For $n = 0, 1, 2, \ldots$, the coefficients can be represented by

$$a_{2k} = (-)^k \frac{n!a_0}{2^{2k}k!(n + k)!}. \qquad (8.141)$$

From now on, we will use the compact notation

$$(-)^k \equiv (-1)^k. \qquad (8.142)$$

Setting $a_0 = 1/2^n n!$, we obtain the conventional definition of a *Bessel function of the first kind*:

$$J_n(x) = \left(\frac{x}{2}\right)^n \left[1 - \frac{(x/2)^2}{2!(n + 2)!} + \frac{(x/2)^4}{4!(n + 4)!} - \cdots \right]. \qquad (8.143)$$

The first three Bessel functions are plotted in Figure 8.5. Their general behavior can be characterized as damped oscillation, qualitatively similar to that in Figure 8.4.

Bessel functions can be generalized for noninteger index ν, as follows:

$$J_\nu(x) = \left(\frac{x}{2}\right)^\nu \left[1 - \frac{(x/2)^2}{2!\Gamma(\nu + 3)} + \frac{(x/2)^4}{4!\Gamma(\nu + 5)} - \cdots \right]. \qquad (8.144)$$

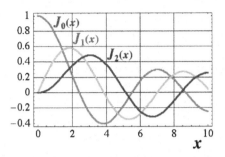

Figure 8.5 Bessel functions of the first kind $J_n(x)$.

The general solution of Bessel's differential equation for noninteger ν is given by

$$y(x) = c_1 J_\nu(x) + c_2 J_{-\nu}(x). \tag{8.145}$$

For integer n, however,

$$J_{-n}(x) = (-)^n J_n(x), \tag{8.146}$$

so that $J_{-n}(x)$ is *not* a linearly independent solution. Following the strategy used in Eq. (8.158), we can construct a second solution by defining

$$Y_\nu(x) = \frac{\cos \nu\pi \, J_\nu(x) - J_{-\nu}(x)}{\sin \nu\pi}. \tag{8.147}$$

In the limit as ν approaches an integer n, we obtain

$$Y_n(x) = \frac{1}{\pi} \left[\frac{\partial}{\partial n} J_n(x) - (-1)^n \frac{\partial}{\partial n} J_{-n}(x) \right]. \tag{8.148}$$

This defines a *Bessel function of the second kind* (sometimes called a Neumann function and written N_n). The computational details are horrible, but fortunately, mathematicians have worked them all out for us and these functions have been extensively tabulated. Figure 8.6 shows the first three functions $Y_n(x)$.

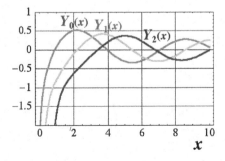

Figure 8.6 Bessel functions of the second kind $Y_n(x)$.

The limiting behavior of the $Y_n(x)$ as $x \to 0$ is apparent from the leading term $J_n(x) \approx (x/2)^n$. Using the definition (8.148) we find

$$Y_n(x) = \frac{2}{\pi} \ln\left(\frac{x}{2}\right) J_n(x) + \text{complicated series.} \tag{8.149}$$

Figure 8.6 shows the logarithmic singularities as $x \to 0$.

8.9 Second Solution

We have encountered two cases in which one solution of a second-order differential equation is relatively straightforward, but the second solution is more obscure. There is a systematic procedure for determining the second solution once the first is known.

Recall that two functions $y_1(x)$ and $y_2(x)$ are *linearly independent* if the relation

$$c_1 y_1(x) + c_2 y_2(x) = 0 \tag{8.150}$$

can only be fulfilled when $c_1 = c_2 = 0$. A test for linear independence is that the *Wronskian*, W, of the two functions is not equal to zero:

$$W(y_1, y_2) \equiv y_1(x)y_2'(x) - y_1'(x)y_2(x) \neq 0. \tag{8.151}$$

If we did have $W = 0$, then $d \ln y_1/dx = d \ln y_2/dx$, which would imply that $y_2 = \text{const } y_1$, thus negating linear independence.

Assume that $y_1(x)$ and $y_2(x)$ are linearly independent solutions to the second-order differential equation (8.131). We can then write

$$\begin{aligned} y_1''(x) + p(x)y_1'(x) + q(x)y_1(x) &= 0, \\ y_2''(x) + p(x)y_2'(x) + q(x)y_2(x) &= 0. \end{aligned} \tag{8.152}$$

Multiplying the first equation by $y_2(x)$, the second by $y_1(x)$ and subtracting, we find

$$\left[y_1(x)y_2''(x) - y_2(x)y_1''(x)\right] + \left[y_1(x)y_2'(x) - y_2(x)y_1'(x)\right] p(x) = 0. \tag{8.153}$$

The second bracket is recognized as the Wronskian $W(y_1, y_2)$, while the first bracket equals dW/dx. Thus

$$\frac{dW}{dx} + W(x)p(x) = 0. \tag{8.154}$$

Separation of variables gives

$$\frac{dW}{W} = -p(x)dx \implies \ln W(x) = -\int p(x)dx + \text{const.} \tag{8.155}$$

Thus

$$W(x) = \text{const } e^{-\int p(x)dx}. \tag{8.156}$$

But

$$W(x) = y_1 y_2' - y_1' y_2 = y_1^2 \frac{d}{dx}\left(\frac{y_2}{y_1}\right),$$ (8.157)

which can be solved for y_2 to give

$$y_2(x) = y_1(x) \int \frac{e^{-\int p(x)dx}}{[y_1(x)]^2} dx.$$ (8.158)

For example, suppose we know one solution $y_1(x) = \sin kx$ for $y''(x)+k^2 y(x) = 0$. Noting that $p(x) = 0$, we can find the second solution

$$y_2(x) = \sin kx \int \frac{dx}{\sin^2 kx} = \sin kx \left(-\frac{1}{k}\cot kx\right) = \text{const} \times \cos kx.$$ (8.159)

Again, for the differential equation $(D - r)^2 y = 0$ or, more explicitly,

$$y''(x) - 2ry'(x) + r^2 y(x) = 0,$$ (8.160)

there is one obvious solution, $y_1(x) = e^{rx}$. Noting that $p(x) = -2r$, the second solution follows from

$$y_2(x) = e^{rx} \int \frac{e^{2rx}}{(e^{rx})^2} dx = x\,e^{rx}.$$ (8.161)

The second solution of Bessel's equation (8.135) for integer n can be found from (8.158) with $p(x) = x^{-1}$:

$$y_2(x) = J_n(x) \int \frac{e^{-\int x^{-1}dx}}{[J_n(x)]^2} dx = J_n(x) \int \frac{dx}{x[J_n(x)]^2}.$$ (8.162)

An expansion for $Y_n(x)$ can be obtained after calculating a power series for $[J_n(x)]^{-2}$.

8.10 Eigenvalue Problems

An *operator*, designated here by $\mathcal{O}$, represents an action which transforms one function into another function. For the case of functions of a single variable:

$$\mathcal{O}f(x) = g(x).$$ (8.163)

We have already encountered the differential operator $D = d/dx$ in Section 8.3. In a sense, an operator is a generalization of the concept of a *function*, which transforms one number into another. For the Schrödinger equation in Eq. (8.101), one can define the *Hamiltonian operator* by

$$H = -\frac{\hbar}{2m}\frac{d^2}{dx^2} + V(x).$$ (8.164)

The Schrödinger equation can then be written more compactly as

$$H\psi(x) = E\psi(x),$$
(8.165)

where E is the energy of the quantum system. As we have seen in Section 8.6, the boundary conditions of the problem determine a set of allowed energies E_n and corresponding wavefunctions $\psi_n(x)$. These are known as *eigenvalues* and *eigenfunctions*, respectively. In pure English these are called "characteristic values" and "characteristic functions," respectively. The corresponding German terms are *eigenwert* and *eigenfunktion*. Current usage employs a hybrid of the English and German words, namely *eigenvalue* and *eigenfunction*.

All operators in quantum mechanics which correspond to observable quantities must be *Hermitian*, such that

$$\int f^*(x)Hg(x)dx = \left(\int g^*(x)Hf(x)dx \right)^*,$$
(8.166)

where $f(x)$ and $g(x)$ obey the same analyticity and boundary conditions that are imposed on the eigenfunctions.

The allowed set of eigenvalues and eigenfunctions obtained for some Schrödingier equation can be summarized symbolically by

$$H\psi_n(x) = E_n\psi_n(x).$$
(8.167)

For another solution, we can write

$$H\psi_m(x) = E_m\psi_m(x).$$
(8.168)

Now multiply the first equation by ψ_m^* and the complex conjugate of the second equation by $\psi_n(x)$. Then subtract the two expressions and integrate over x. The result can be written

$$\int \psi_m^*(x)H\psi_n(x)dx - \left(\int \psi_n^*(x)H\psi_m(x)dx \right)^* = (E_n - E_m^*)\int \psi_m^*(x)\psi_n(x)dx.$$
(8.169)

But by the presumed Hermitian property of H, the left-hand side equals zero. Thus

$$(E_n - E_m^*)\int \psi_m^*(x)\psi_n(x)dx = 0.$$
(8.170)

If $m = n$, the second factor becomes $\int |\psi_n(x)|^2 dx$, which is nonzero. Therefore the first factor must equal zero, meaning that

$$E_n^* = E_n.$$
(8.171)

Thus the energy eigenvalues must be real numbers, which is reasonable, given that they are measurable physical quantities.

Next consider the case when $E_m \neq E_n$. Then it must be the second factor in Eq. (8.170) that equals zero:

$$\int \psi_m^*(x)\psi_n(x)dx = 0 \quad \text{when } E_m \neq E_n. \tag{8.172}$$

Thus eigenfunctions belonging to unequal eigenvalues are *orthogonal*. There remains the case that for some $m \neq n$, $E_m = E_n$. The eigenfunctions ψ_m and ψ_n are then said to be *degenerate*. But it is always possible to construct linear combinations of degenerate eigenfunctions so that the orthogonality relation still applies. If the eigenfunctions are, in addition, all normalized, we obtain the compact *orthonormality* condition

$$\int \psi_m^*(x)\psi_n(x)dx = \delta_{mn}. \tag{8.173}$$

The eigenfunctions $\{\psi_n(x)\}$ then constitute an *orthonormal set*.

Problem 8.10.1.   Show that a linear combination of degenerate eigenfunctions is itself an eigenfunction with the same energy eigenvalue.

9 Matrix Algebra

Thus far, we have been doing algebra involving numbers and functions. It is also possible to apply the operations of algebra to more general types of mathematical entities. In this chapter, we will deal with *matrices*, which are *ordered arrays* of numbers or functions. For example, a matrix which we designate by the symbol $\mathbb{A}$ can represent a collection of quantities arrayed as follows:

$$\mathbb{A} = \begin{bmatrix} a_{11} & a_{12} & a_{13} & \cdots & a_{1n} \\ a_{21} & a_{22} & a_{23} & \cdots & a_{2n} \\ a_{31} & a_{32} & a_{33} & \cdots & a_{3n} \\ \vdots & \vdots & \vdots & & \vdots \\ a_{n1} & a_{n2} & a_{n3} & \cdots & a_{nn} \end{bmatrix}. \tag{9.1}$$

The subscripts i and j on the *matrix elements* a_{ij} label the *rows* and *columns*, respectively. The matrix $\mathbb{A}$ shown above is an $n \times n$ *square matrix*, with n rows and n columns. We will also make use of $n \times 1$ *column matrices* or *column vectors* such as

$$\mathbb{x} = \begin{bmatrix} x_1 \\ x_2 \\ x_3 \\ \vdots \\ x_n \end{bmatrix} \tag{9.2}$$

and $1 \times n$ *row matrices* or *row vectors* such as

$$\tilde{\mathbb{x}} = [\, x_1 \quad x_2 \quad x_3 \quad \cdots \quad x_n\,]. \tag{9.3}$$

Where do matrices come from? Suppose we have a set of n simultaneous relations, each involving n quantities $x_1, x_2, x_3, \ldots, x_n$:

$$a_{11}x_1 + a_{12}x_2 + a_{13}x_3 + \cdots + a_{1n}x_n = y_1,$$
$$a_{21}x_1 + a_{22}x_2 + a_{23}x_3 + \cdots + a_{2n}x_n = y_2,$$
$$a_{31}x_1 + a_{32}x_2 + a_{33}x_3 + \cdots + a_{3n}x_n = y_3,$$
$$\vdots$$
$$a_{n1}x_1 + a_{n2}x_2 + a_{n3}x_3 + \cdots + a_{nn}x_n = y_n. \tag{9.4}$$

This set of n relations can be represented symbolically by a single matrix equation

$$\mathbb{A}\mathbb{x} = \mathbb{y}, \tag{9.5}$$

Guide to Essential Math 2e. http://dx.doi.org/10.1016/B978-0-12-407163-6.00009-6

where $\mathbb{A}$ is the $n \times n$ matrix (9.1), while X and Y are $n \times 1$ column vectors, such as (9.2).

9.1   Matrix Multiplication

Comparing (9.4) with (9.5), it is seen that matrix multiplication implies the following rule involving their component elements:

$$\sum_{k=1}^{n} a_{ik}x_k = y_i, \quad i = 1, 2, \ldots, n. \tag{9.6}$$

Note that summation over identical adjacent indices k results in their mutual "annihilation." Suppose the quantities y_i in Eq. (9.4) are themselves determined by n simultaneous relations

$$b_{11}y_1 + b_{12}y_2 + b_{13}y_3 + \cdots + b_{1n}y_n = z_1,$$
$$b_{21}y_1 + b_{22}y_2 + b_{23}y_3 + \cdots + b_{2n}y_n = z_2,$$
$$b_{31}y_1 + b_{32}y_2 + b_{33}y_3 + \cdots + b_{3n}y_n = z_3,$$
$$\vdots$$
$$b_{n1}y_1 + b_{n2}y_2 + b_{n3}y_3 + \cdots + b_{nn}y_n = z_n. \tag{9.7}$$

The combined results of Eqs. (9.4) and (9.7), equivalent to eliminating $y_1, y_2, \ldots, y_n$ between the two sets of equations, can be written

$$c_{11}x_1 + c_{12}x_2 + c_{13}x_3 + \cdots + c_{1n}x_n = z_1,$$
$$c_{21}x_1 + c_{22}x_2 + c_{23}x_3 + \cdots + c_{2n}x_n = z_2,$$
$$c_{31}x_1 + c_{32}x_2 + c_{33}x_3 + \cdots + c_{3n}x_n = z_3,$$
$$\vdots$$
$$c_{n1}x_1 + c_{n2}x_2 + c_{n3}x_3 + \cdots + c_{nn}x_n = z_n. \tag{9.8}$$

We can write the same equations in matrix notation:

$$\mathbb{A}\mathrm{X} = \mathrm{Y}, \quad \mathbb{B}\mathrm{Y} = \mathrm{Z} \implies \mathbb{B}\mathbb{A}\mathrm{X} = \mathrm{Z} \implies \mathbb{C}\mathrm{X} = \mathrm{Z}. \tag{9.9}$$

Evidently, $\mathbb{C}$ can be represented as a *matrix product*:

$$\mathbb{C} = \mathbb{B}\mathbb{A}. \tag{9.10}$$

An element of the product matrix is constructed by summation over two sets of matrix elements in the following pattern:

$$\sum_{k=1}^{n} b_{ik}a_{kj} = c_{ij}. \tag{9.11}$$

The diagram below shows schematically how the ijth element is constructed from the sum of products of elements from the ith row of the first matrix and the jth column of the second:

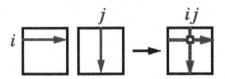

The three 2×2 Pauli spin matrices

$$\sigma_1 = \begin{bmatrix} 0 & 1 \\ 1 & 0 \end{bmatrix}, \quad \sigma_2 = \begin{bmatrix} 0 & -i \\ i & 0 \end{bmatrix}, \quad \sigma_3 = \begin{bmatrix} 1 & 0 \\ 0 & -1 \end{bmatrix} \tag{9.12}$$

will provide computationally simple examples to illustrate many of the properties of matrices. They are themselves of major significance in applications to quantum mechanics and geometry.

The most dramatic contrast between multiplication of matrices and multiplication of numbers is that matrix multiplication can be *noncommutative*, meaning that it is not necessarily true that

$$AB = BA. \tag{9.13}$$

As a simple illustration, consider products of Pauli spin matrices: We find

$$\sigma_1 \sigma_2 = \begin{bmatrix} i & 0 \\ 0 & -i \end{bmatrix} = i\sigma_3 \quad \text{but } \sigma_2 \sigma_1 = \begin{bmatrix} 0 & i \\ -i & 0 \end{bmatrix} = i\sigma_3, \tag{9.14}$$

also

$$\sigma_1 \sigma_3 = \begin{bmatrix} 0 & -1 \\ 1 & 0 \end{bmatrix} = -i\sigma_2 \quad \text{but } \sigma_3 \sigma_1 = \begin{bmatrix} 0 & 1 \\ -1 & 0 \end{bmatrix} = i\sigma_2. \tag{9.15}$$

Matrix multiplication remains *associative*, however, so that

$$A(BC) = (AB)C = ABC. \tag{9.16}$$

In matrix multiplication, the product of an $n \times m$ matrix and an $m \times p$ matrix is an $n \times p$ matrix. Two matrices cannot be multiplied unless their adjacent dimensions—p in the above example—match. As we have seen above, square matrix multiplying a column vector gives another column vector ($[n \times n][n \times 1] \rightarrow [n \times 1]$). The product of a row vector and a column vector is an ordinary number (in a sense, a 1×1 matrix). For example,

$$\tilde{X}Y = x_1 y_1 + x_2 y_2 + \cdots + x_n y_n. \tag{9.17}$$

Problem 9.1.1. Calculate and compare the matrix products $\sigma_2\sigma_3$ and $\sigma_3\sigma_2$.

Problem 9.1.2. The *commutator* of two matrices is defined by

$$[\mathbb{A}, \mathbb{B}] \equiv \mathbb{A}\mathbb{B} - \mathbb{B}\mathbb{A}$$

and the *anticommutator* by

$$\{\mathbb{A}, \mathbb{B}\} \equiv \mathbb{A}\mathbb{B} + \mathbb{B}\mathbb{A}.$$

Calculate the commutators and anticommutators for each pair of Pauli matrices.

9.2 Further Properties of Matrices

Following are a few hints on how to manipulate indices in matrix elements. It is most important to recognize that any index that is summed over is a *dummy index*. The result is independent of what we call it. Thus

$$\sum_{i=1}^{n} a_i = \sum_{j=1}^{n} a_j = \sum_{i=k}^{n} a_k, \text{ etc.} \tag{9.18}$$

Secondly, it is advisable to use different indices when a product of summations occurs in an expression. For example,

$$\sum_{i=1}^{n} a_i \sum_{i=1}^{n} b_i \text{ is better written as } \sum_{i=1}^{n} a_i \sum_{j=1}^{n} b_j.$$

This becomes mandatory if we reexpress it as a double summation

$$\sum_{i=1}^{n}\sum_{j=1}^{n} a_i b_j.$$

Multiplication of a matrix $\mathbb{A}$ by a constant c is equivalent to multiplying each a_{ij} by c. Two matrices *of the same dimension* can be added element by element. By combination of these two operations, the matrix elements of $\mathbb{C} = k_1\mathbb{A} + k_2\mathbb{B}$ are given by $c_{ij} = k_1 a_{ij} + k_2 b_{ij}$.

The *null matrix* has all its elements equal to zero:

$$\mathbb{O} \equiv \begin{bmatrix} 0 & 0 & \cdots & 0 \\ 0 & 0 & \cdots & 0 \\ \vdots & \vdots & & \vdots \\ 0 & 0 & \cdots & 0 \end{bmatrix}. \tag{9.19}$$

As expected,

$$\mathbb{A}\mathbb{O} = \mathbb{O}\mathbb{A} = \mathbb{O}. \tag{9.20}$$

A *diagonal matrix* has only nonvanishing elements along the main diagonal, for example

$$\Lambda = \begin{bmatrix} \lambda_1 & 0 & 0 & \cdots & 0 \\ 0 & \lambda_2 & 0 & \cdots & 0 \\ 0 & 0 & \lambda_3 & \cdots & 0 \\ \vdots & \vdots & \vdots & & \vdots \\ 0 & 0 & 0 & \cdots & \lambda_n \end{bmatrix}. \tag{9.21}$$

Its elements can be written in terms of the Kronecker delta:

$$\Lambda_{ij} = \lambda_i \delta ij. \tag{9.22}$$

A diagonal matrix is sometimes represented in a form such as

$$\Lambda = \mathrm{diag}\{\lambda_1, \lambda_2, \ldots, \lambda_n\}. \tag{9.23}$$

A special case is the *unit* or *identity matrix*, diagonal with all elements equal to 1:

$$\mathbb{I} = \begin{bmatrix} 1 & 0 & 0 & \cdots & 0 \\ 0 & 1 & 0 & \cdots & 0 \\ 0 & 0 & 1 & \cdots & 0 \\ \vdots & \vdots & \vdots & & \vdots \\ 0 & 0 & 0 & \cdots & 1 \end{bmatrix}. \tag{9.24}$$

Clearly,

$$\mathbb{I}_{ij} = \delta_{ij}. \tag{9.25}$$

As expected, for an arbitrary matrix $\mathbb{A}$:

$$\mathbb{A}\mathbb{I} = \mathbb{I}\mathbb{A} = \mathbb{A}. \tag{9.26}$$

9.3 Determinants

Determinants, an important adjunct to matrices, can be introduced as a geometrical construct. Consider the parallelogram shown in Figure 9.1, with one vertex at the origin $(0, 0)$ and the other three at (x_1, y_1), (x_2, y_2), and $(x_1 + x_2, y_1 + y_2)$. Using Pythagoras' theorem, the two sides a, b and the diagonal c have the lengths

$$a = \sqrt{x_1^2 + y_1^2}, \quad b = \sqrt{x_2^2 + y_2^2}, \quad c = \sqrt{(x_2 - x_1)^2 + (y_2 - y_1)^2}. \tag{9.27}$$

The area of the parallelogram is given by

$$\pm A = ab \sin \theta, \tag{9.28}$$

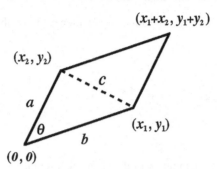

Figure 9.1 Area of parallelepiped equals determinant $\begin{vmatrix} x_1 & x_2 \\ y_1 & y_2 \end{vmatrix}$.

where θ is the angle between sides a and b. The $\pm$ sign is determined by the relative orientation of (x_1, y_1) and (x_2, y_2). Also, by the law of cosines,

$$c^2 = a^2 + b^2 - 2ab\cos\theta. \tag{9.29}$$

Eliminating θ between Eqs. (9.28) and (9.29), we find, after some lengthy algebra, that

$$\pm A = x_1 y_2 - y_1 x_2. \tag{9.30}$$

(If you know about the cross product of vectors, this follows directly from $A = \mathbf{a} \times \mathbf{b} = ab\sin\theta = x_1 y_2 - y_1 x_2$.) This combination of variables has the form of a *determinant*, written

$$\begin{vmatrix} x_1 & x_2 \\ y_1 & y_2 \end{vmatrix} = x_1 y_2 - y_1 x_2. \tag{9.31}$$

In general for a 2×2 matrix $\mathbb{M}$

$$\det \mathbb{M} = \begin{vmatrix} m_{11} & m_{12} \\ m_{21} & m_{22} \end{vmatrix} = m_{11}m_{22} - m_{12}m_{21}. \tag{9.32}$$

The three-dimensional analog of a parallelogram is a parallelepiped, with all six faces being parallelograms. As shown in Figure 9.2, the parallelepiped is oriented between the origin $(0, 0, 0)$ and the point $(x, y, z) = (x_1 + x_2 + x_3, y_1 + y_2 + y_3, z_1 + z_2 + z_3)$, which is the reflection of the origin through the plane containing the points (x_1, y_1, z_1), (x_2, y_2, z_2), and (x_3, y_3, z_3). You can figure out, using some algebra and trigonometry, that the volume is given by

$$\pm V = \begin{vmatrix} x_1 & x_2 & x_3 \\ y_1 & y_2 & y_3 \\ z_1 & z_2 & z_3 \end{vmatrix} = x_1 y_2 z_3 + y_1 z_2 x_3 + z_1 x_2 y_3 - x_3 y_2 z_1 - y_3 z_2 x_1 - z_3 x_2 y_1. \tag{9.33}$$

[Using vector analysis, $\pm V = \mathbf{a} \times \mathbf{b} \cdot \mathbf{c}$, where $\mathbf{a}, \mathbf{b}, \mathbf{c}$ are the vectors from the origin to (x_1, y_1, z_1), (x_2, y_2, z_2), (x_3, y_3, z_3), respectively.]

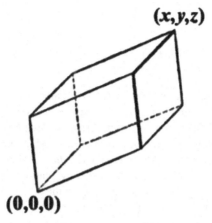

Figure 9.2 Volume of parallelepiped equals determinant $\begin{vmatrix} x_1 & x_2 & x_3 \\ y_1 & y_2 & y_3 \\ z_1 & z_2 & z_3 \end{vmatrix}$.

It might be conjectured that an $n \times n$-determinant represents the hypervolume of an n-dimensional hyperparallelepiped.

In general, a 3×3 determinant is given by

$$\det M = \begin{vmatrix} m_{11} & m_{12} & m_{13} \\ m_{21} & m_{22} & m_{23} \\ m_{31} & m_{32} & m_{33} \end{vmatrix} = m_{11}m_{22}m_{33} + m_{12}m_{23}m_{31} + m_{13}m_{21}m_{32}$$

$$- m_{13}m_{22}m_{31} - m_{12}m_{21}m_{33} - m_{11}m_{23}m_{32}. \tag{9.34}$$

A 2×2 determinant can be evaluated by summing over products of elements along the two diagonals, northwest-southeast minus northeast-southwest:

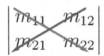

Similarly for a 3×3 determinant:

where the first two columns are duplicated on the right. There is no simple graphical method for 4×4 or larger determinants. An $n \times n$ determinant is defined more generally by

$$\det \mathbb{M} = \sum_{p=1}^{n!} (-1)^p \, \mathcal{P}[m_{1i} m_{2j} m_{3k} \cdots], \qquad (9.35)$$

where $\mathcal{P}$ is a *permutation operator* which runs over all $n!$ possible permutations of the indices $i, j, k, \ldots$ The permutation label p is even or odd, depending on the number of binary interchanges of the second indices necessary to obtain $m_{1i} \, m_{2j} \, m_{3k} \cdots$, starting from its order on the main diagonal: $m_{11} \, m_{22} \, m_{33} \cdots$. Many math books show further reductions of determinants involving *minors* and *cofactors*, but this is no longer necessary with readily available computer programs to evaluate determinants. An important property of determinants, which is easy to verify in the 2×2 and 3×3 cases, is that if any two rows or columns of a determinant are interchanged, the value of the determinant is multiplied by -1. As a corollary, if any two rows or two columns are identical, the determinant equals zero.

The determinant of a product of two matrices, in either order, equals the product of their determinants. More generally for a product of three or more matrices, in any cyclic order,

$$\det(\mathbb{ABC}) = \det(\mathbb{BCA}) = \det(\mathbb{CAB}) = \det \mathbb{A} \det \mathbb{B} \det \mathbb{C}. \qquad (9.36)$$

Problem 9.3.1.   Find the volume of a unit cube coincident with the coordinate axes by evaluating a 3×3 determinant.

Problem 9.3.2.   As a more challenging variant, calculate the volume of a rotated unit cube with one vertex standing on the origin.

9.4 Matrix Inverse

The *inverse* of a matrix $\mathbb{M}$, designated $\mathbb{M}^{-1}$, satisfies the matrix equation

$$\mathbb{M}\mathbb{M}^{-1} = \mathbb{M}^{-1}\mathbb{M} = \mathbb{I}. \qquad (9.37)$$

For the 2×2 matrix

$$\mathbb{M} = \begin{bmatrix} m_{11} & m_{22} \\ m_{21} & m_{22} \end{bmatrix}.$$

The inverse is given by

$$\mathbb{M}^{-1} = \frac{1}{m_{11}m_{22} - m_{12}m_{21}} \begin{bmatrix} m_{22} & -m_{12} \\ -m_{21} & m_{11} \end{bmatrix}. \qquad (9.38)$$

For matrices of larger dimension, the inverses can be readily evaluated by computer programs. Note that the denominator in (9.38) equals the *determinant* of the matrix $\mathbb{M}$.

In order for the inverse $\mathbb{M}^{-1}$ to exist, the determinant of a matrix must *not* be equal to zero. Consequently, a matrix with determinant equal to zero is termed *singular*. A matrix with $\det \mathbb{M} = 1$ is called *unimodular*.

The inverse of a product of matrices equals the product of inverses *in reversed order*. For example,

$$(\mathbb{ABC})^{-1} = \mathbb{C}^{-1}\mathbb{B}^{-1}\mathbb{A}^{-1}. \tag{9.39}$$

You can easily prove this by multiplying by $\mathbb{ABC}$.

The inverse matrix can be used to solve a series of simultaneous linear equations, such as (9.4). Supposing the y_i are known quantities while the x_i are unknowns, multiply the matrix equation (9.5) by $\mathbb{A}^{-1}$. This gives

$$\mathbb{A}^{-1}\mathbb{A}\mathbb{x} = \mathbb{E}\mathbb{x} = \mathbb{x} = \mathbb{A}^{-1}\mathbb{y}. \tag{9.40}$$

With the elements of $\mathbb{A}^{-1}$ and $\mathbb{y}$ known, the column vector $\mathbb{x}$, hence its elements x_i can be determined. The solutions are given explicitly by *Cramer's rule*:

$$x_i = \frac{D_i}{D}, \tag{9.41}$$

where D is the determinant of the matrix $\mathbb{A}$:

$$D = \begin{vmatrix} a_{11} & a_{12} & a_{13} & \cdots & a_{1n} \\ a_{21} & a_{22} & a_{23} & \cdots & a_{2n} \\ a_{31} & a_{32} & a_{33} & \cdots & a_{3n} \\ \vdots & \vdots & \vdots & & \vdots \\ a_{n1} & a_{n2} & a_{n3} & \cdots & a_{nn} \end{vmatrix} \tag{9.42}$$

and D_i is obtained from D by replacing the ith column by the column vector $\mathbb{y}$:

$$D_i = \begin{vmatrix} a_{11} & a_{12} & \cdots & y_1 & \cdots & a_{1n} \\ a_{21} & a_{22} & \cdots & y_2 & \cdots & a_{2n} \\ a_{31} & a_{32} & \cdots & y_3 & \cdots & a_{3n} \\ \vdots & \vdots & & \vdots & & \vdots \\ a_{n1} & a_{n2} & \cdots & y_n & \cdots & a_{nn} \end{vmatrix}. \tag{9.43}$$

A set of *homogeneous linear equations*

$$\begin{aligned} a_{11}x_1 + a_{12}x_2 + a_{13}x_3 + \cdots + a_{1n}x_n &= 0, \\ a_{21}x_1 + a_{22}x_2 + a_{23}x_3 + \cdots + a_{2n}x_n &= 0, \\ a_{31}x_1 + a_{32}x_2 + a_{33}x_3 + \cdots + a_{3n}x_n &= 0, \\ &\vdots \\ a_{n1}x_1 + a_{n2}x_2 + a_{n3}x_3 + \cdots + a_{nn}x_n &= 0, \end{aligned} \tag{9.44}$$

always has the *trivial solution* $x_1 = x_2 = \cdots = x_n = 0$. A *necessary* condition for a nontrivial solution to exist is that $\det \mathbb{A} = 0$. (This is not a *sufficient* condition, however. The trivial solution might still be the only one.)

Problem 9.4.1. Find the inverses of the three Pauli matrices, σ_1, σ_2, and σ_3.

Problem 9.4.2. Using Cramer's rule, solve the set of simultaneous linear equations

$$x + 2y + 3z = 4,$$
$$2x + 3y + z = 5,$$
$$3x + y + 2z = 6.$$

9.5 Wronskian Determinant

A set of n functions $\{f_1(x), f_2(x), \ldots, f_n(x)\}$ is said to be *linearly independent* if vanishing of the linear combination

$$c_1 f_1(x) + c_2 f_2(x) + \cdots + c_n f_n(x) = 0 \tag{9.45}$$

can only be achieved with the "trivial" solution

$$c_1 = c_2 = \cdots = c_n = 0.$$

A criterion for linear independence can be obtained by constructing a set of n simultaneous equations involving (9.45) along with its 1st, 2nd, ..., $(n-1)$st derivatives:

$$c_1 f_1(x) + c_2 f_2(x) + \cdots + c_n f_n(x) = 0,$$
$$c_1 f_1'(x) + c_2 f_2'(x) + \cdots + c_n f_n'(x) = 0,$$
$$c_1 f_1''(x) + c_2 f_2''(x) + \cdots + c_n f_n''(x) = 0,$$
$$\vdots$$
$$c_1 f_1^{(n-1)}(x) + c_2 f_2^{(n-1)}(x) + \cdots + c_n f_n^{(n-1)}(x) = 0. \tag{9.46}$$

A trivial solution, hence linear independence, is guaranteed if the *Wronskian determinant* is nonvanishing, i.e.

$$W[f_1, f_2, \ldots, f_n] \equiv \begin{vmatrix} f_1(x) & f_2(x) & \cdots & f_n(x) \\ f_1'(x) & f_2'(x) & & f_n'(x) \\ f_1''(x) & f_2''(x) & \cdots & f_n''(x) \\ \vdots & \vdots & & \vdots \\ f_1^{(n-1)}(x) & f_2^{(n-1)}(x) & \cdots & f_n^{(n-1)}(x) \end{vmatrix} \neq 0. \tag{9.47}$$

You can show, for example, that the set $\{\cos x, e^{ix}\}$ is linearly independent, while the set $\{\cos x, \sin x, e^{ix}\}$ is not.

Problem 9.5.1. Test the pair of functions $\{\cos x, e^{ix}\}$ for linear independence.

Problem 9.5.2. Similarly test the set $\{\cos x, \sin x, e^{ix}\}$.

9.6 Special Matrices

The *transpose* of a matrix, designated $\tilde{M}$ or M^T, is obtained by interchanging its rows and columns or, alternatively, by reflecting all the matrix elements through the main diagonal:

$$M \rightarrow \tilde{M} \quad \text{when } m_{ij} \rightarrow m_{ji} \text{ all } i, j. \tag{9.48}$$

A matrix equal to its transpose, $M = \tilde{M}$, is called *symmetric*. Two examples of symmetric matrices are

$$\sigma_1 = \begin{bmatrix} 0 & 1 \\ 1 & 0 \end{bmatrix} \quad \text{and} \quad \sigma_3 = \begin{bmatrix} 1 & 0 \\ 0 & -1 \end{bmatrix}. \tag{9.49}$$

If $M = -\tilde{M}$, the matrix is *skew-symmetric*, for example

$$\sigma_2 = \begin{bmatrix} 0 & -i \\ i & 0 \end{bmatrix}. \tag{9.50}$$

A matrix is *orthogonal* if its transpose equals its inverse: $\tilde{R} = R^{-1}$. A 2×2 unimodular orthogonal matrix—also known as a *special orthogonal matrix*—can be expressed in the form

$$R = \begin{bmatrix} \cos\theta & -\sin\theta \\ \sin\theta & \cos\theta \end{bmatrix}. \tag{9.51}$$

The totality of such two-dimensional matrices is known as the *special orthogonal group*, designated SO(2). The rotation of a Cartesian coordinate system in a plane, such that

$$\begin{aligned} x' &= x\cos\theta - y\sin\theta, \\ y' &= x\sin\theta + y\cos\theta, \end{aligned} \tag{9.52}$$

can be compactly represented by the matrix equation

$$\begin{bmatrix} x' \\ y' \end{bmatrix} = R \begin{bmatrix} x \\ y \end{bmatrix}. \tag{9.53}$$

Since R is orthogonal, $R\tilde{R} = I$, which leads to the invariance relation

$$x'^2 + y'^2 = x^2 + y^2. \tag{9.54}$$

As a general principle, a linear transformation preserves length if and only if its matrix is orthogonal.

The *Hermitian conjugate* of a matrix, $M^\dagger$, is obtained by transposition accompanied by complex conjugation:

$$M \rightarrow M^\dagger \quad \text{when } m_{ij} \rightarrow m_{ji}^* \text{ all } i, j. \tag{9.55}$$

A matrix is *Hermitian* or *self-adjoint* if $\mathbb{H}^\dagger = \mathbb{H}$. The matrices σ_1, σ_2, and σ_3 introduced above are all Hermitian. The Hermitian conjugate of a product equals the product of conjugates in reverse order:

$$(\mathbb{AB})^\dagger = \mathbb{B}^\dagger \mathbb{A}^\dagger \tag{9.56}$$

analogous to the inverse of a product. The same ordering is true for the transpose of a product. Also, it should be clear that a second Hermitian conjugation returns a matrix to its original form:

$$(\mathbb{H}^\dagger)^\dagger = \mathbb{H}. \tag{9.57}$$

The analogous effect of double application is also true for the inverse and the transpose. A matrix is *unitary* if its Hermitian conjugate equals its inverse: $\mathbb{U}^\dagger = \mathbb{U}^{-1}$. The set of 2×2 unimodular unitary matrices constitutes the *special unitary group* SU(2). Such matrices can be parametrized by

$$\begin{bmatrix} a & b \\ -b^* & a^* \end{bmatrix} \quad \text{with } |a|^2 + |b|^2 = 1 \tag{9.58}$$

or by

$$\begin{bmatrix} e^{i\phi}\cos\theta & -e^{i\phi}\sin\theta \\ e^{-i\phi}\sin\theta & e^{-i\phi}\cos\theta \end{bmatrix}. \tag{9.59}$$

The SU(2) matrix group is of significance in the physics of spin-$\frac{1}{2}$ particles.

9.7 Similarity Transformations

A matrix $\mathbb{M}$ is said to undergo a *similarity transformation* to $\mathbb{M}'$ if

$$\mathbb{M}' = \mathbb{TMT}^{-1}, \tag{9.60}$$

where the *transformation matrix* $\mathbb{T}$ is nonsingular. (The transformation is alternatively written $\mathbb{M}' = \mathbb{T}^{-1}\mathbb{MT}$.) When the matrix $\mathbb{R}$ is orthogonal, we have an *orthogonal transformation*: $\mathbb{M}' = \mathbb{RM\tilde{R}}$. When the transformation matrix is unitary, we have a *unitary transformation*: $\mathbb{M}' = \mathbb{UMU}^\dagger$. All similarity transformations preserve the form of matrix equations. Suppose

$$\mathbb{AB} = \mathbb{C}.$$

Premultiplying by $\mathbb{T}$ and postmultiplying by $\mathbb{T}^{-1}$, we have

$$\mathbb{TABT}^{-1} = \mathbb{TCT}^{-1}.$$

Inserting $\mathbb{I}$ in the form of $\mathbb{T}^{-1}\mathbb{T}$ between $\mathbb{A}$ and $\mathbb{B}$:

$$\mathbb{TAT}^{-1}\mathbb{TBT}^{-1} = \mathbb{TCT}^{-1}.$$

From the definition of primed matrices in Eq. (9.60), we conclude

$$\mathbb{A}\mathbb{B} = \mathbb{C} \implies \mathbb{A}'\mathbb{B}' = \mathbb{C}'. \tag{9.61}$$

This is what we mean by the *form* of a matrix relation being preserved under a similarity transformation. The determinant of a matrix is also invariant under a similarity transformation, since

$$\det \mathbb{M}' = \det(\mathbb{T}\mathbb{M}\mathbb{T}^{-1}) = \det(\mathbb{M}\mathbb{T}^{-1}\mathbb{T}) = \det \mathbb{M}. \tag{9.62}$$

9.8 Matrix Eigenvalue Problems

One important application of similarity transformations is to reduce a matrix to diagonal form. This is particularly relevant in quantum mechanics, when the matrix is Hermitian and the transformation unitary. Consider the relation

$$\mathbb{U}^\dagger \mathbb{H} \mathbb{U} = \Lambda, \tag{9.63}$$

where Λ is a diagonal matrix, such as (9.21). Premultiplying by $\mathbb{U}$, this becomes

$$\mathbb{H}\mathbb{U} = \mathbb{U}\Lambda. \tag{9.64}$$

Expressed in terms of matrix elements:

$$\sum_k H_{ik} U_{kj} = \sum_k U_{ik} \Lambda_{kj} = U_{ij}\lambda_j, \tag{9.65}$$

recalling that the elements of the diagonal matrix are given by $\Lambda_{kj} = \lambda_j \delta_{kj}$ and noting that only the term with $k = j$ will survive the summation over k. The unitary matrix $\mathbb{U}$ can be pictured as composed of an array of column vectors $\mathbb{x}^{(j)}$, such that $x_i^{(j)} = U_{ij}$, like this:

$$\mathbb{U} = \begin{bmatrix} x_1^{(1)} \\ x_2^{(1)} \\ x_3^{(1)} \\ \vdots \\ x_n^{(1)} \end{bmatrix} \begin{bmatrix} x_1^{(2)} \\ x_2^{(2)} \\ x_3^{(2)} \\ \vdots \\ x_n^{(2)} \end{bmatrix} \begin{bmatrix} x_1^{(3)} \\ x_2^{(3)} \\ x_3^{(3)} \\ \vdots \\ x_n^{(3)} \end{bmatrix} \begin{matrix} \vdots \\ \vdots \\ \vdots \\ \vdots \\ \vdots \end{matrix} \begin{bmatrix} x_1^{(n)} \\ x_2^{(n)} \\ x_3^{(n)} \\ \vdots \\ x_n^{(n)} \end{bmatrix}. \tag{9.66}$$

Accordingly Eq. (9.64) can be written as a set of equations

$$\mathbb{H}\mathbb{x}^{(j)} = \lambda_j \mathbb{x}^{(j)}, \quad j = 1, 2, \ldots, n, \tag{9.67}$$

This is an instance of an *eigenvalue equation*. In general, a matrix $\mathbb{A}$ operating on a vector $\mathbb{X}$ will produce another vector $\mathbb{Y}$, as shown in Eq. (9.5). For certain very special vectors $\mathbb{x}_n$, the matrix multiplication miraculously reproduces the original vector multiplied by a constant λ_n, so that

$$\mathbb{A}\mathbb{x}_n = \lambda_n \mathbb{x}_n. \tag{9.68}$$

Eigenvalue problems are most frequently encountered in quantum mechanics. The differential equation for the particle-in-a-box, treated in Section 8.6, represents another type of eigenvalue problem. There, the boundary conditions restricted the allowed energy values to the discrete set E_n, enumerated in Eq. (8.111). These are consequently called *energy eigenvalues*.

The eigenvalues of a Hermitian matrix are real numbers. This follows by taking the Hermitian conjugate of Eq. (9.63):

$$(\mathbb{U}^\dagger \mathbb{H} \mathbb{U})^\dagger = \mathbb{U}^\dagger \mathbb{H}^\dagger \mathbb{U} = \mathbf{\Lambda}^\dagger. \tag{9.69}$$

Since $\mathbb{H}^\dagger = \mathbb{H}$, by its Hermitian property, we conclude that

$$\mathbf{\Lambda}^\dagger = \mathbf{\Lambda} \; \Rightarrow \; \lambda_j^* = \lambda_j \quad \text{all } j. \tag{9.70}$$

Hermitian eigenvalues often represent physically observable quantities, consistent with their values being real numbers.

The eigenvalues and eigenvectors can be found by solving the set of simultaneous linear equations represented by (9.67):

$$
\begin{aligned}
H_{11}x_1 + H_{12}x_2 + H_{13}x_3 + \cdots + H_{1n}x_n &= \lambda x_1, \\
H_{21}x_1 + H_{22}x_2 + H_{23}x_3 + \cdots + H_{2n}x_n &= \lambda x_2, \\
H_{31}x_1 + H_{32}x_2 + H_{33}x_3 + \cdots + H_{3n}x_n &= \lambda x_3, \\
&\vdots \\
H_{n1}x_1 + H_{n2}x_2 + H_{n3}x_3 + \cdots + H_{nn}x_n &= \lambda x_n.
\end{aligned}
\tag{9.71}
$$

This reduces to a set of homogeneous equations:

$$
\begin{aligned}
(H_{11} - \lambda)x_1 + H_{12}x_2 + H_{13}x_3 + \cdots + H_{1n}x_n &= 0, \\
H_{21}x_1 + (H_{22} - \lambda)x_2 + H_{23}x_3 + \cdots + H_{2n}x_n &= 0, \\
H_{31}x_1 + H_{32}x_2 + (H_{33} - \lambda)x_3 + \cdots + H_{3n}x_n &= 0, \\
&\vdots \\
H_{n1}x_1 + H_{n2}x_2 + H_{n3}x_3 + \cdots + (H_{nn} - \lambda)x_n &= 0.
\end{aligned}
\tag{9.72}
$$

A necessary condition for a nontrivial solution is the vanishing of the determinant:

$$
\begin{vmatrix}
H_{11} - \lambda & H_{12} & H_{13} & \cdots & H_{1n} \\
H_{21} & H_{22} - \lambda & H_{23} & \cdots & H_{2n} \\
H_{31} & H_{32} & H_{33} - \lambda & \cdots & H_{3n} \\
\vdots & \vdots & \vdots & & \vdots \\
H_{n1} & H_{n2} & H_{n3} & \cdots & H_{nn} - \lambda
\end{vmatrix} = 0,
\tag{9.73}
$$

this is known as the *secular equation* and can be solved for n roots $\lambda_1, \lambda_2, \ldots, \lambda_n$.

It is a general result that the eigenvectors of two unequal eigenvalues are orthogonal. To prove this, consider two different eigensolutions of a matrix $\mathbb{A}$:

$$\mathbb{A}\mathbb{X}_n = \lambda_n \mathbb{X}_n, \quad \mathbb{A}\mathbb{X}_m = \lambda_m \mathbb{X}_m, \quad \text{with} \quad \lambda_m \neq \lambda_n. \tag{9.74}$$

Now, take the Hermitian conjugate of the n equation, recalling that $\mathbb{A}$ is Hermitian ($\mathbb{A}^\dagger = \mathbb{A}$) and λ_n is real ($\lambda_n^* = \lambda_n$). Thus

$$(\mathbb{A}\mathbb{X}_n)^\dagger = \mathbb{X}_n^\dagger \mathbb{A}. \tag{9.75}$$

Now postmultiply the last equation by $\mathbb{X}_m$, premultiply the m equation by $\mathbb{X}_n^\dagger$, and subtract the two. The result is

$$(\lambda_m - \lambda_n)\mathbb{X}_m^\dagger \mathbb{X}_n = 0. \tag{9.76}$$

If $\lambda_m \neq \lambda_n$, then $\mathbb{X}_m$ and $\mathbb{X}_n$ are orthogonal:

$$\mathbb{X}_m^\dagger \mathbb{X}_n = 0. \tag{9.77}$$

When $\lambda_m = \lambda_n$, although $m \neq n$, the proof fails. The two eigenvectors $\mathbb{X}_m$ and $\mathbb{X}_n$ are said to be *degenerate*. It is still possible to find a linear combination of $\mathbb{X}_m$ and $\mathbb{X}_n$ so that the orthogonality relation Eq. (9.77) still applies. If, in addition, all the eigenvectors are normalized, meaning that

$$\mathbb{X}_n^\dagger \mathbb{X}_n = 1 \quad \text{all } n \tag{9.78}$$

then the set of eigenvectors $\{\mathbb{X}_n\}$ constitutes an *orthonormal set* satisfying the compact relation

$$\mathbb{X}_m^\dagger \mathbb{X}_n = \delta_{m,n} \tag{9.79}$$

analogous to the relation for orthonormalized eigenfunctions.

In quantum mechanics there is a very fundamental connection between matrices and integrals involving operators and their eigenfunctions. A matrix we denote as $\mathbb{H}$ is defined such that its matrix elements correspond to integrals over an operator H and its eigenfunctions $\psi_n(x)$, constructed as follows:

$$H_{mn} = \int \psi_m^*(x) H \psi_n(x) dx. \tag{9.80}$$

The two original formulations of quantum mechanics were Heisenberg's *matrix mechanics* (1925), based on representation of observables by noncommuting matrices and Schrödinger's *wave mechanics* (1926), based on operators and differential equations. It was deduced soon afterward by Schrödinger and by Dirac that the two formulations were equivalent representations of the same underlying physical theory, a key connection being the equivalence between matrices and operators demonstrated above.

Problem 9.8.1. Find the eigenvalues and normalized eigenvectors for each of the three Pauli matrices, σ_1, σ_2, and σ_3.

Problem 9.8.2. Find the eigenvalues and normalized eigenvectors of the matrix

$$\begin{bmatrix} 1 & 0 & -i \\ 0 & 1 & 0 \\ i & 0 & 1 \end{bmatrix}.$$

Problem 9.8.3. Show that the matrix representation of a Hermitian operator, as defined in Eq. (8.166), corresponds to a Hermitian matrix.

9.9   Diagonalization of Matrices

A matrix $\mathbb{M}$ is *diagonalizable* if there exists a similarity transformation of the form

$$A^{-1}\mathbb{M}A = \Lambda = \text{diag}\{\lambda_1, \lambda_2, \ldots, \lambda_n\}. \tag{9.81}$$

All Hermitian, symmetric, unitary, and orthogonal matrices are diagonalizable, as is any $n \times n$-matrix whose n eigenvalues are distinct. The process of diagonalization is essentially equivalent to determination of the eigenvalues of a matrix, which are given by the diagonal elements λ_n.

The *trace* of a matrix is defined as the sum of its diagonal elements:

$$\text{tr}\,\mathbb{M} = \sum_n M_{nn}. \tag{9.82}$$

This can be shown to be equal to the sum of its eigenvalues. Since

$$\mathbb{M} = A\Lambda A^{-1}, \tag{9.83}$$

we can write

$$\sum_n M_{nn} = \sum_{n,k,j} A_{nk}\Lambda_{kj}A_{jn}^{-1} = \sum_k \lambda_k \sum_n A_{kn}^{-1}A_{nk} = \sum_k \lambda_k, \tag{9.84}$$

noting that $A^{-1}A = \mathbb{I}$. Therefore

$$\text{tr}\,\mathbb{M} = \sum_n \lambda_n. \tag{9.85}$$

Problem 9.1.1.   Find similarity transformations which diagonalize the Pauli matrices σ_1 and σ_2.

9.10   Four-Vectors and Minkowski Spacetime

Suppose that at $t = 0$ a light flashes at the origin, creating a spherical wave propagating outward at the speed of light c. The locus of the wavefront will be given by

$$x^2 + y^2 + z^2 = c^2t^2. \tag{9.86}$$

According to Einstein's Special Theory of Relativity, the wave will retain its spherical appearance to every observer, even one moving at a significant fraction of the speed of light. This can be expressed mathematically as the invariance of the differential element

$$ds^2 = c^2\,dt^2 - dx^2 - dy^2 - dz^2 \tag{9.87}$$

known as the *spacetime interval*. Equation (9.87) has a form suggestive of Pythagoras' theorem in four dimensions. It was fashionable in the early years of the 20th century to define an imaginary time variable $x_4 = ict$, which together with the space variables $x_1 = x, x_2 = y$, and $x_3 = z$ forms a pseudo-Euclidean four-dimensional space with interval given by

$$ds^2 = dx_1^2 + dx_2^2 + dx_3^2 + dx_4^2. \tag{9.88}$$

This contrived Euclidean geometry doesn't change the reality that time is fundamentally very different from a spatial variable. It is current practice to accept the differing signs in the spacetime interval and define a *real* time variable $x^0 = ct$, in terms of which

$$ds^2 = (dx^0)^2 - (dx^1)^2 - (dx^2)^2 - (dx^3)^2. \tag{9.89}$$

The corresponding geometrical structure is known as *Minkowski spacetime*. The form we have written, described as having the *signature* $\{+ - --\}$, is preferred by elementary-particle physicists. People working in General Relativity write instead $ds^2 = -(dx^0)^2 + (dx^1)^2 + (dx^2)^2 + (dx^3)^2$, with signature $\{- + ++\}$.

The spacetime variables are the components of a Minkowski *four-vector*, which can be thought of as a column vector

$$x^\mu = \begin{bmatrix} x^0 \\ x^1 \\ x^2 \\ x^3 \end{bmatrix} \tag{9.90}$$

with its differential analog

$$dx^\mu = \begin{bmatrix} dx^0 \\ dx^1 \\ dx^2 \\ dx^3 \end{bmatrix}. \tag{9.91}$$

Specifically, these are *contravariant* four-vectors, with their component labels written as *superscripts*. The spacetime interval (9.89) can be represented as a scalar product if we define associated *covariant* four-vectors as the row matrices

$$x_\mu = [\, x_0 \quad x_1 \quad x_2 \quad x_3 \,] \quad \text{and} \quad dx_\mu = [\, dx_0 \quad dx_1 \quad dx_2 \quad dx_3 \,], \tag{9.92}$$

with the component indices written as *subscripts*. A matrix product can then be written:

$$ds^2 = dx_\mu dx^\mu = [\, dx_0 \quad dx_1 \quad dx_2 \quad dx_3 \,] \begin{bmatrix} dx^0 \\ dx^1 \\ dx^2 \\ dx^3 \end{bmatrix}$$

$$= dx_0\, dx^0 + dx_1\, dx^1 + dx_2\, dx^2 + dx_3\, dx^3. \tag{9.93}$$

This accords with (9.89) provided that the covariant components x_μ are given by

$$x_0 = x^0 = ct, \quad x_1 = -x^1 = -x, \quad x_2 = -x^2 = -y, \quad x_3 = -x^3 = -z. \tag{9.94}$$

It is convenient to introduce the *Einstein summation convention* for products of covariant and contravariant vectors, whereby

$$a_\mu b^\mu \equiv \sum_{\mu=0}^{3} a_\mu b^\mu. \tag{9.95}$$

Any term containing the same Greek covariant and contravariant indices is understood to be summed over that index. This applies even to *tensors*, objects with multiple indices. For example, a valid tensor equation might read

$$A_\nu^{\mu\lambda} B_\lambda^\kappa = C_\nu^{\mu\kappa}. \tag{9.96}$$

The equation applies for all values of the indices which are *not* summed over. The index λ summed from 0 to 3 is said to be *contracted*. Usually, the summation convention for Latin indices implies a sum just from 1 to 3, for example

$$a_k b^k \equiv a_1 b^1 + a_2 b^2 + a_3 b^3 = -a^1 b^1 - a^2 b^2 - a^3 b^3 = -\mathbf{a} \cdot \mathbf{b}. \tag{9.97}$$

A four-dimensional scalar product can alternatively be written

$$a_\mu b^\mu = a_0 b^0 - \mathbf{a} \cdot \mathbf{b}. \tag{9.98}$$

Covariant and contravariant vectors can be interconverted with use of the *metric tensor* $\eta_{\mu\nu}$, given by

$$\eta_{\mu\nu} = \eta^{\mu\nu} = \begin{bmatrix} 1 & 0 & 0 & 0 \\ 0 & -1 & 0 & 0 \\ 0 & 0 & -1 & 0 \\ 0 & 0 & 0 & -1 \end{bmatrix}. \tag{9.99}$$

For example,

$$a_\mu = \eta_{\mu\nu} a^\nu, \quad a^\mu = \eta^{\mu\nu} a_\nu. \tag{9.100}$$

The spacetime interval takes the form

$$ds^2 = \eta_{\mu\nu} \, dx^\mu \, dx^\nu. \tag{9.101}$$

In General Relativity, the metric tensor $g_{\mu\nu}$ is determined by the curvature of spacetime and the interval generalizes to

$$ds^2 = g_{\mu\nu} \, dx^\mu \, dx^\nu, \tag{9.102}$$

where $g_{\mu\nu}$ might have some nonvanishing off-diagonal elements. In flat spacetime (in the absence of curvature), this reduces to Special Relativity with $g_{\mu\nu} = \eta_{\mu\nu}$.

The energy and momentum of a particle in relativistic mechanics can also be represented as components of a four-vector p^μ with

$$p^0 = E/c, \quad p^1 = p_x, \quad p^2 = p_y, \quad p^3 = p_z, \tag{9.103}$$

and correspondingly

$$p_0 = E/c, \quad p_1 = -p_x, \quad p_2 = -p_y, \quad p_3 = -p_z. \tag{9.104}$$

The scalar product is an invariant quantity

$$p_\mu p^\mu = m^2 c^2, \tag{9.105}$$

where m is the rest mass of the particle. Written out explicitly, this gives the relativistic energy-momentum relation:

$$E^2 - p^2 c^2 = m^2 c^4. \tag{9.106}$$

In the special case of a particle at rest $\mathbf{p} = 0$, we obtain Einstein's famous mass-energy equation $E = mc^2$. The alternative root $E = -mc^2$ is now understood to pertain to the corresponding *antiparticle*. For a particle with zero rest mass, such as the photon, we obtain $p = E/c$. Recalling that $\lambda v = c$, this last four-vector relation is consistent with both the Planck and de Broglie formulas: $E = hv$ and $p = h/\lambda$.

10 Group Theory

10.1 Introduction

In this chapter, we will provide neither proofs nor detailed applications of group theory for either mathematics or scientific subjects. It will suffice to acquaint you with some of the motivation, terminology, and possible uses of this highly developed and widely applied branch of mathematics. Group theory deals with collections of entities which can be transformed among themselves by some appropriate operation. For example, the integers constitute a group (containing an infinite number of members) which can be transformed into one another by the operation of addition (which includes subtraction). Likewise, under the operation of multiplication, the four complex numbers $\{i, -1, -i, +1\}$ can be recycled among themselves. Two versions of the Yin and Yang symbol (Figure 10.1) can be turned into one another by reversing the colors black and white or by rotating the figure by 180°. Palindromes are words or phrases which read the same in either direction—like "RADAR" and Napoleon's lament, "Able was I ere I saw Elba." And, of course, everyone has marveled at the beautiful sixfold symmetry of snowflakes. Interpersonal behavior can develop symmetry, as in "tit for tat." The algebraic expression $x^2 + y^2$ is unchanged in value when x and y are interchanged. Likewise, the forms of Maxwell's equations in free space (in the appropriate units) are preserved when the fields $\mathbf{E}$ and $-\mathbf{B}$ are interchanged. Group-theoretical relationships are often revealed by *patterns* among the members of the collection. For example, in elementary particle theory, the systematic arrangement of masses, and spins of hadrons (particles subject to the strong interaction) eventually led to the quark model for baryons and mesons.

Figure 10.1 Yin-Yang symbol and its color inverse. The original could then be retrieved by a 180° rotation. (For interpretation of the references to color in this figure legend, the reader is referred to the web version of this book.)

Guide to Essential Math 2e. http://dx.doi.org/10.1016/B978-0-12-407163-6.00010-2

Symmetry in a technical sense implies that certain things remain invariant even when they are subject to some type of transformation. In fact, some of the fundamental laws of physics can be based on exact or approximate invariance of systems under certain real or abstract symmetry operations. In several important instances, an invariance or a symmetry implies a conservation law, a general principle known as *Noether's theorem*. For example, the equations of mechanics appear to be invariant to an advance or retardation in the time variable. From this invariance, the conservation of energy can be deduced. Similarly, invariance with respect to translation and rotation in space implies conservation of linear and angular momentum, respectively. In quantum mechanics, the invariance of the Schrödinger equation with respect to a phase factor $e^{i\alpha}$ in the wavefunction $\Psi(\mathbf{r}, t)$ implies the conservation of electric charge. More generally, localized invariance with respect to a phase factor $e^{i\chi(\mathbf{r},t)}$ implies the existence of the quantized electromagnetic field. Such principles can be formalized as *gauge field theories*, which provide the basic structure of the Standard Model for electromagnetic, weak, and strong interactions.

The strong nuclear force is insensitive to the distinction between neutrons and protons. These can be treated as alternative states of a single particle called a *nucleon*, differing in *isotopic spin* or *isospin*. It is found, for example, that the nuclei ^{3}H and ^{3}He have similar energy-level spectra. Isospin is, however, only an approximate symmetry. It is "broken" by electromagnetic interactions, since protons have electric charge while neutrons do not. Broken symmetry is a central theme in fundamental physics. An open question is how our Universe evolved to break the symmetry between matter and antimatter, so that it is now dominated by matter.

10.2   Symmetry Operations

As a more concrete and elementary introduction to group theory, consider the symmetry operations which transform an equilateral triangle into an indistinguishable copy of itself. These are shown in Figure 10.2, with the vertices labeled as 1, 2, and 3. A group always contains an *identity element*, designated E, which represents the default operation of "doing nothing." A positive (counterclockwise) rotation by an angle of $2\pi/3$ is designated C_3 and the corresponding clockwise rotation is designated $\overline{C}_3$. Reflections (or 180° rotations) through the three vertices are designated σ_1, σ_2, and σ_3. These symmetry operations are represented in Figure 10.2 using colored triangles.

The definitive property of a group is that successive application of two operations is equivalent to some single operation. For example,

$$\sigma_1 C_3 = \sigma_3, \tag{10.1}$$

where the operation on the *right* is understood to be performed first. For the same two operations in reversed order, we find

$$C_3 \sigma_1 = \sigma_2. \tag{10.2}$$

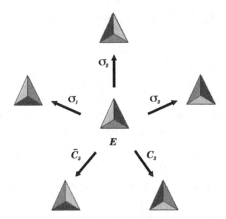

Figure 10.2 Symmetry operations on an equilateral triangle, shown with the aid of colored areas. (For interpretation of the references to color in this figure legend, the reader is referred to the web version of this book.)

Thus, group elements do *not*, in general, commute

$$G_1 G_2 \neq G_2 G_1 \tag{10.3}$$

although they *may* commute, as do C_3 and $\overline{C}_3$. (A group in which all products of elements commute is known as an *Abelian group*.) The algebra of the group can be summarized by the following 6×6 *group multiplication table*:

RIGHT	LEFT	E	C_3	$\overline{C}_3$	σ_1	σ_2	σ_3
E		E	C_3	$\overline{C}_3$	σ_1	σ_2	σ_3
C_3		C_3	$\overline{C}_3$	E	σ_3	σ_1	σ_2
$\overline{C}_3$		$\overline{C}_3$	E	C_3	σ_2	σ_3	σ_1
σ_1		σ_1	σ_2	σ_3	E	C_3	$\overline{C}_3$
σ_2		σ_2	σ_3	σ_1	$\overline{C}_3$	E	C_3
σ_3		σ_3	σ_1	σ_2	C_3	$\overline{C}_3$	E

Note that each operation appears exactly once, and only once, in each row and in each column. The group describing the symmetry operations on an equilateral triangle is designated in molecular theory as $\mathcal{C}_{3v}$ has precisely the same structure as the $3! = 6$ possible permutations of three objects. The latter is known as the *symmetric group* of order 3, designated $\mathcal{S}_3$. These symmetry and permutation groups are said to be *isomorphous*—their abstract properties are identical although they apply to completely different sorts of objects.

10.3   Mathematical Theory of Groups

A *group* is defined as a set of h abstract elements $\mathcal{G} \equiv \{G_1, G_2, \ldots, G_h\}$ together with a rule for combination of these elements, which we usually refer to as a *product*. The mathematical properties of groups are entirely independent of any physical attributes that can be ascribed to their elements or rule of combination. For concreteness, however, we will usually apply the concepts of group theory to sets of geometrical symmetry operations. The elements of a group must fulfill the following four conditions:

1. The product of any two elements of the group gives another element of the group. That is, $G_i G_j = G_k$ with $G_k \in \mathcal{G}$.
2. Group multiplication obeys an associative law, $G_i(G_j G_k) = (G_i G_j)G_k \equiv G_i G_j G_k$.
3. There exists an *identity element* E such that $E G_i = G_i E = G_i$ for all G_i.
4. Every element G_i has a unique inverse G_i^{-1}, such that $G_i G_i^{-1} = G_i^{-1} G_i = E$ with $G_i^{-1} \in \mathcal{G}$.

You can verify each of these conditions in the multiplication table for the group $\mathcal{C}_{3v}$ shown in the previous section. The number of elements h is called the *order* of the group. Thus, $\mathcal{C}_{3v}$ is a group of order $g = 6$.

A subgroup is a smaller group selected from the elements of a given group, which itself fulfills all the properties of a group. For example, the set $\{E, C_3, \overline{C}_3\}$ is itself a group of order 3. Likewise, each of the sets $\{E, \sigma_k\}$ for $k = 1, 2, 3$ is a subgroup of order 2. Finally the identity element E is the trivial case of a subgroup of order 1. It can be proven that the order of any subgroup must be a divisor of the order of the group g.

Every element of a group, say X, can be subjected to a set of similarity transformations using all the elements of the group, G_k with $k = 1, 2, \ldots, g$, whereby

$$G_k^{-1} X G_k = X_k. \tag{10.4}$$

The element X, along with all the elements X_k thus generated (there may be duplicates) are said to belong to the same *class*. The group $\mathcal{C}_{3v}$ has three classes: $\{\sigma_1, \sigma_2, \sigma_3\}$, $\{C_3, \overline{C}_3\}$, and $\{E\}$. Evidently, for symmetry groups, the members of each class perform similar geometric operations, but oriented differently in space. The identity element, E, is always in a class by itself. The class structure of this group can be designated $\{E, 2C_3, 3\sigma_v\}$. Again, the number of members or each class must be a divisor of the order of the group g.

10.4   Representations of Groups

A set of quantities which obeys the group multiplication table is called a *representation* of the group. Because of the possible noncommutativity of group elements, simple numbers are not always adequate to represent groups; we must often use matrices. The group $\mathcal{C}_{3v}$ has three *irreducible representations*, or IRs, which cannot be broken down into simpler representations. A trivial, but nonetheless important, representation of every group is the *totally symmetric representation*, in which each group element is represented by 1. The multiplication table then simply reiterates that $1 \times 1 = 1$. For S_3 this is called the A_1 representation:

$$A_1: \quad E = 1, \; C_3 = 1, \; \overline{C}_3 = 1, \; \sigma_1 = 1, \; \sigma_2 = 1, \; \sigma_3 = 1. \tag{10.5}$$

A slightly less trivial representation is A_2:

$$A_2: \quad E = 1, \; C_3 = 1, \; \overline{C}_3 = 1, \; \sigma_1 = -1, \; \sigma_2 = -1, \; \sigma_3 = -1. \tag{10.6}$$

Much more exciting is the E representation, which requires 2×2 matrices:

$$
E = \begin{bmatrix} 1 & 0 \\ 0 & 1 \end{bmatrix}, \quad
C_3 = \begin{bmatrix} -1/2 & -\sqrt{3}/2 \\ \sqrt{3}/2 & -1/2 \end{bmatrix},
$$
$$
\overline{C}_3 = \begin{bmatrix} -1/2 & \sqrt{3}/2 \\ -\sqrt{3}/2 & -1/2 \end{bmatrix}, \quad
\sigma_1 = \begin{bmatrix} -1 & 0 \\ 0 & 1 \end{bmatrix},
$$
$$
\sigma_2 = \begin{bmatrix} 1/2 & -\sqrt{3}/2 \\ -\sqrt{3}/2 & -1/2 \end{bmatrix}, \quad
\sigma_3 = \begin{bmatrix} 1/2 & \sqrt{3}/2 \\ \sqrt{3}/2 & -1/2 \end{bmatrix}. \tag{10.7}
$$

We can also find the following three-dimensional representation of the group, which we designate by Γ:

$$
E = \begin{bmatrix} 1 & 0 & 0 \\ 0 & 1 & 0 \\ 0 & 0 & 1 \end{bmatrix}, \quad
C_3 = \begin{bmatrix} 0 & 1 & 0 \\ 0 & 0 & 1 \\ 1 & 0 & 0 \end{bmatrix},
$$
$$
\overline{C}_3 = \begin{bmatrix} 0 & 0 & 1 \\ 1 & 0 & 0 \\ 0 & 1 & 0 \end{bmatrix}, \quad
\sigma_1 = \begin{bmatrix} 1 & 0 & 0 \\ 0 & 0 & 1 \\ 0 & 1 & 0 \end{bmatrix},
$$
$$
\sigma_2 = \begin{bmatrix} 0 & 0 & 1 \\ 0 & 1 & 0 \\ 1 & 0 & 0 \end{bmatrix}, \quad
\sigma_3 = \begin{bmatrix} 0 & 1 & 0 \\ 1 & 0 & 0 \\ 0 & 0 & 1 \end{bmatrix}. \tag{10.8}
$$

This is however a *reducible representation* of the group. By this we mean that there exists a similarity transformation, $S^{-1}\Gamma(G)S$, which when applied to each of the matrices in (10.8) reduces them all to *block diagonal* form. For the Γ representation, the matrix

$$
S = \begin{bmatrix} 1/\sqrt{3} & 1/\sqrt{6} & -1/\sqrt{2} \\ 1/\sqrt{3} & -2/\sqrt{6} & 0 \\ 1/\sqrt{3} & 1/\sqrt{6} & 1/\sqrt{2} \end{bmatrix} \tag{10.9}
$$

gives a reduction into a block diagonal form

$$
S^{-1}\Gamma(G)S = \begin{bmatrix} \square & 0 & 0 \\ 0 & \square & \square \\ 0 & \square & \square \end{bmatrix} \tag{10.10}
$$

in which upper box stands for a one-dimensional representation, while the block of four boxes stands for a two-dimensional representation. In this case, these are, respectively, the A_1 and E representations. This reduction of the Γ representation can be expressed symbolically as

$$\Gamma = A_1 \oplus E. \tag{10.11}$$

A representation which cannot be further transformed into lower-dimensional representations is called an *irreducible representation*. We state without proof that the number of irreducible representations of a group is equal to the number of classes. Another important theorem states that the sum of the squares of the dimensionalities of the irreducible representations of a group adds up to the order of the group. Thus, for C_{3v}, we find $1^2 + 1^2 + 2^2 = 6$. The general result is

$$\sum_r d_r^2 = g, \tag{10.12}$$

where d_r is the dimension of the irreducible representation r.

We have usually referred to the operations of a group generically as *multiplications*. But, as mentioned earlier, addition can also be considered a group operation.

Problem 10.3.1. Verify the following cute matrix representation which mirrors the addition of two numbers $x + y$:

$$\begin{bmatrix} 1 & x \\ 0 & 1 \end{bmatrix}\begin{bmatrix} 1 & y \\ 0 & 1 \end{bmatrix} = \begin{bmatrix} 1 & x+y \\ 0 & 1 \end{bmatrix}.$$

Problem 10.3.2. Work out explicitly the reduction of the Γ representation for the group element C_3 into block diagonal form.

10.5 Group Characters

The *trace* or *character* of a matrix is defined as the sum of the elements along the main diagonal:

$$\chi(\mathbb{M}) = \sum_{k=1}^{g} M_{kk}. \tag{10.13}$$

For many purposes, it suffices to know just the characters of a matrix representation of a group, rather than the complete matrices. For example, the characters for the E representation of C_{3v} in Eq. (10.7) are given by

$$\chi(E) = 2, \quad \chi(C_2) = -1, \quad \chi(\overline{C}_2) = -1, \quad \chi(\sigma_1) = \chi(\sigma_2) = \chi(\sigma_3) = 0. \tag{10.14}$$

Note that the characters for all operations in the same class are equal. Thus, the preceding equation can be abbreviated to

$$\chi(E) = 2, \quad \chi(C_3) = -1, \quad \chi(\sigma_v) = 0. \tag{10.15}$$

For one-dimensional representations, such as A_1 and A_2, the characters are equal to the matrices themselves, so Eqs. (10.5) and (10.6) can be directly read as character tables.

The characters of a group have an interesting orthonormality property:

$$\sum_G \chi_r(G)\chi_s(G) = g\delta_{rs}, \tag{10.16}$$

where the sum is over all the elements of the group, with r and s labeling two representations of the group. The last formula can, in fact, be derived from what is known as the *Wonderful Orthogonality Theorem*, which we state without proof:

$$\sum_G \Gamma^r(G)_{mn}^* \Gamma^s(G)_{m'n'} = \frac{g}{d_r}\delta_{rs}\delta_{mm'}\delta_{nn'}. \tag{10.17}$$

Here $\Gamma^r(G)_{mn}$ means the mnth matrix element of the representation r, with dimension d_r, of the group element G.

For many purposes, particularly in applications to molecules, the essential information about a symmetry group is summarized in its character table. The character table for C_{3v} is shown here:

C_{3v}	E	$2C_3$	$3\sigma_v$		
A_1	1	1	1	z	$x^2 + y^2, z^2$
A_2	1	1	-1	R_z	
E	2	-1	0	$(x, y)(R_x, R_y)$	$(x^2 - y^2, xy)(xz, yz)$

The last two columns show how the Cartesian coordinates x, y, z, combinations of Cartesian coordinates and rotations R_x, R_y, R_z transform under the operations of the group. This is the only character table we will display explicitly. Character tables for all the relevant symmetry groups are given in many textbooks on group theory, quantum chemistry, and spectroscopy.

Problem 10.5.1. Show that characters of the Γ representation of C_{3v} obey the relation

$$\chi_\Gamma(G) = \chi_{A_1}(G) + \chi_E(G)$$

for every member of the group. This is a standard method for determining the composition of a reducible representation.

10.6   Group Theory in Quantum Mechanics

When a molecule belongs to the symmetry of a group $\mathcal{G}$, this means that each member of the group commutes with the molecular Hamiltonian

$$[G_k, H] = 0 \quad \text{for } k = 1, 2, \ldots, g. \tag{10.18}$$

We now treat the group elements G_k as operators on wavefunctions or orbitals ψ, which are generally functions in three dimensions. Commuting operators can have simultaneous eigenfunctions. A representation of the group of dimension d means that there must exist a set of d degenerate eigenfunctions of H, which we designate by ψ_{nk} that transform among themselves in accord with the corresponding d-dimensional matrix representation of the group $\mathcal{G}$. For example, if the eigenvalue E_n is d-fold degenerate, the vanishing commutators imply that, for

$$G_k H \psi_{nk} = H(G_k \psi_{nk}) = E_n(G_k \psi_{nk}), \quad k = 1, 2, \ldots, d. \tag{10.19}$$

Thus each $G\psi_{nk}$ is also an eigenfunction of H with the same eigenvalue E_n and must therefore be representable as a linear combination of the eigenfunctions ψ_{nk} More precisely, the eigenfunctions transform among themselves according to the relation

$$G_i \psi_{nk} = \sum_{m=1}^{d} \Gamma(G_i)_{km} \psi_{nm}, \tag{10.20}$$

where $\Gamma(G_i)_{km}$ means, in this case, the $\{k, m\}$ element of the matrix representing the operator G_i. The character of the identity operation E immediately shows the degeneracy of the eigenvalues of that symmetry. The C_{3v} character table reveals that molecules of that same symmetry (for example, ammonia, NH_3) can have only nondegenerate and twofold degenerate energy levels.

10.7 Molecular Symmetry Operations

Much information about molecules can be deduced from their symmetry group, even without quantum-mechanical computations. Specifically, we are dealing with *point groups*, so called because at least one point in the molecule remains fixed under all the symmetry operations of the group. Following is an enumeration of the types of symmetry operations that are needed to describe molecular configurations:

1. E: The identity transformation meaning do nothing (from the German Einheit, meaning unity.)
2. C_n: Clockwise rotation by an angle of $2!/n$ radians (n is an integer). The axis for which n is maximum is called the principal axis.
3. σ: Reflection through a plane (Spiegel is German for mirror).
4. σ_h: Horizontal reflection plane, one passing through the origin and perpendicular to the principal axis.
5. σ_v: Vertical reflection plane, one passing through the origin and containing the principal axis.
6. σ_d: Diagonal or dihedral reflection in a plane through the origin and containing the principal axis. Similar to σ_v except that it also bisects the angles between two C_2 axes perpendicular to the principal axis.
7. i: Inversion through the origin. In Cartesian coordinates, the transformation $(x, y, z) \rightarrow (-x, -y, -z)$. Irreducible representations under this symmetry operation are classified as g (even) or u (odd).

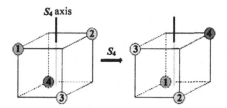

Figure 10.3 Pictorial representation of the operation S4.

8. S_n: An improper rotation or rotation-reflection axis. Clockwise rotation through an angle of $2!/n$ radians followed by a reflection in the plane perpendicular to the axis of rotation. Also known as an alternating axis of symmetry. Note that S_1 is equivalent to σ_h and S_2 is equivalent to i.

Improper rotations can be visualized as shown in Figure 10.3.

Every molecule can be characterized as belonging to some group of symmetry operations from the above list, under which it can be transformed into indistinguishable copies of itself. We cannot, however, have arbitrary combinations of symmetry operations. For example, a molecule with a C_n axis of rotation can only have mirror planes which either contain the axis or are perpendicular to it. We have been using the Schoenflies classification scheme, most favored by chemists. (Crystallographers generally use the International or Hermann-Mauguin classification).

A symbol such as C_n actually has a triple meaning in group theory. It can represent a symmetry element, namely an n-fold axis of rotation. It also designates the symmetry operation of rotation by $2\pi/n$ radians about this axis. Finally, in the caligraphic form $\mathcal{C}_n$, it represents the symmetry group containing the elements $\{E, C_n, C_n^2, \ldots, C_n^{n+1}\}$.

Molecular structures illustrating the most important symmetry operations are given in the companion volume: S. M. Blinder, *Introduction to Quantum Mechanics*, Elsevier, 2004, pp. 206-212.

11 Multivariable Calculus

Most physical systems are characterized by more than two quantitative variables. Experience has shown that it is not always possible to change such quantities at will, but that specification of some of them will determine definite values for others. Functional relations involving three or more variables lead us to branches of calculus which make use of *partial derivatives* and *multiple integrals*.

11.1 Partial Derivatives

We have already snuck in the concept of partial differentiation in several instances by evaluating the derivative with respect to x of a function of the form $f(x, y, \ldots)$, while treating $y, \ldots$ as if they were constant quantities. The correct notation for such operations makes use of the "curly dee" symbol ∂, for example

$$\frac{\partial}{\partial x} f(x, y, \ldots) = \frac{\partial f}{\partial x}. \tag{11.1}$$

To begin with, we will consider functions of just two independent variables, such as $z = f(x, y)$. Generalization to more than two variables is usually straightforward. The definitions of partial derivatives are closely analogous to that of the ordinary derivative:

$$\left(\frac{\partial z}{\partial x}\right)_y \equiv \lim_{\Delta x \to 0} \left[\frac{z(x + \Delta x, y) - z(x, y)}{\Delta x}\right] \tag{11.2}$$

and

$$\left(\frac{\partial z}{\partial y}\right)_x \equiv \lim_{\Delta y \to 0} \left[\frac{z(x, y + \Delta y) - z(x, y)}{\Delta y}\right]. \tag{11.3}$$

The subscript y or x denotes the variable which is held *constant*. If if there is no ambiguity, the subscript can be omitted, as in (11.1). Some alternative notations for $(\partial z/\partial x)_y$ are $\partial z/\partial x$, $\partial f/\partial x$, z_x, f_x, and $f^{(1,0)}(x, y)$. As shown in Figure 11.1, a partial derivative such as $(\partial z/\partial x)_y$ can be interpreted geometrically as the instantaneous slope at the point (x, y) of the curve formed by the intersection of the surface $z = z(x, y)$ and the plane $y = $ constant, and analogously for $(\partial z/\partial y)_x$. Partial derivatives can be evaluated by the same rules as for ordinary differentiation, treating all but one variable as constants.

Guide to Essential Math 2e. http://dx.doi.org/10.1016/B978-0-12-407163-6.00011-4

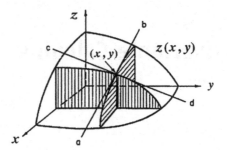

Figure 11.1 Graphical representation of partial derivatives. The curved surface represents $z(x, y)$ in the first quadrant. Vertically and horizontally cross-hatched planes are $x =$ constant and $y =$ constant, respectively. Lines ab and cd are drawn tangent to the surface at point (x, y). The slopes of ab and cd equal $(\partial z/\partial x)_y$ and $(\partial z/\partial y)_x$, respectively.

Products of partial derivatives can be manipulated in the same way as products of ordinary derivatives *provided that the same variables are held constant*. For example,

$$\left(\frac{\partial z}{\partial t}\right)_z \left(\frac{\partial t}{\partial x}\right)_z = \left(\frac{\partial y}{\partial x}\right)_z \tag{11.4}$$

and

$$\left(\frac{\partial y}{\partial x}\right)_z \left(\frac{\partial x}{\partial y}\right)_z = 1 \quad \Rightarrow \quad \left(\frac{\partial x}{\partial y}\right)_z = \frac{1}{(\partial y/\partial x)_z}. \tag{11.5}$$

Since partial derivatives are also functions of the independent variables, they can themselves be differentiated to give second and higher derivatives. These are written, for example, as

$$\frac{\partial^2 z}{\partial x^2} \equiv \frac{\partial}{\partial x}\left(\frac{\partial z}{\partial x}\right) = \left[\frac{\partial}{\partial x}\left(\frac{\partial z}{\partial x}\right)_y\right]_y \tag{11.6}$$

or, in more compact notation, z_{xx}. Also possible are *mixed second derivatives* such as

$$\frac{\partial^2 z}{\partial y \partial x} \equiv \frac{\partial}{\partial y}\left(\frac{\partial z}{\partial x}\right) = \left[\frac{\partial}{\partial y}\left(\frac{\partial z}{\partial x}\right)_y\right]_x. \tag{11.7}$$

When a function and its first derivatives are single valued and continuous, the order of differentiation can be reversed, so that

$$\frac{\partial^2 z}{\partial y \partial x} = \frac{\partial^2 z}{\partial x \partial y} \tag{11.8}$$

or, more compactly, $z_{yx} = z_{xy}$. Higher-order derivatives such as z_{xxx} and z_{xyy} can also be constructed.

Thus far we have considered changes in $z(x, y)$ brought about by changing one variable at a time. The more general case involves *simultaneous* variation of x and y. This could be represented in Figure 11.1 by a slope of the surface $z = f(x, y)$ cut in a direction *not* parallel to either the x- or y-axis. Consider the more general increment

$$\Delta z = z(x + \Delta x, y + \Delta y) - z(x, y). \tag{11.9}$$

Adding and subtracting the quantity $z(x, y + \Delta y)$ and inserting the factors $\Delta x / \Delta x$ and $\Delta y / \Delta y$, we find

$$\Delta z = \left[\frac{z(x + \Delta x, y + \Delta y) - z(x, y + \Delta y)}{\Delta x} \right] \Delta x + $$
$$\left[\frac{z(x, y + \Delta y) - z(x, y)}{\Delta y} \right] \Delta y. \tag{11.10}$$

Passing to the limit $\Delta x \to 0$, $\Delta y \to 0$, the two bracketed quantities approach the partial derivatives (11.2) and (11.3). The remaining increments $\Delta x, \Delta y, \Delta z$ approach the differential quantities dx, dy, dz. The result is the *total differential*:

$$dz = \left(\frac{\partial z}{\partial x} \right)_y dy + \left(\frac{\partial z}{\partial y} \right)_x dx. \tag{11.11}$$

Extension of the total differential to functions of more than two variables is straightforward. For a function of n variables, $f = f(x_1, x_2, \ldots, x_n)$, the total differential is given by

$$df = \left(\frac{\partial f}{\partial x_1} \right) dx_1 + \left(\frac{\partial f}{\partial x_2} \right) dx_1 + \cdots + \left(\frac{\partial f}{\partial x_n} \right) dx_n = \sum_{i=1}^{n} \left(\frac{\partial f}{\partial x_i} \right) dx_i. \tag{11.12}$$

A neat relation among the three partial derivatives involving x, y, and z can be derived from Eq. (11.11). Consider the case when $z = $ constant, so that $dz = 0$. We have then that

$$\left(\frac{\partial z}{\partial x} \right)_y dy + \left(\frac{\partial z}{\partial y} \right)_x dx = 0 \quad \Rightarrow \quad \frac{dy}{dx} = -\frac{(\partial z / \partial x)_y}{(\partial z / \partial y)_x}. \tag{11.13}$$

But the ratio of dy to dx means, in this instance, $(\partial y / \partial x)_z$, since z was constrained to a constant value. Thus we obtain the important identity

$$\left(\frac{\partial y}{\partial x} \right)_z = -\frac{(\partial z / \partial x)_y}{(\partial z / \partial y)_x} \tag{11.14}$$

or in a cyclic symmetrical form

$$\left(\frac{\partial y}{\partial x} \right)_z \left(\frac{\partial z}{\partial y} \right)_x \left(\frac{\partial x}{\partial z} \right)_y = -1. \tag{11.15}$$

As an illustration, suppose we need to evaluate $(\partial V/\partial T)_p$ for one mole of a gas obeying Dieterici's equation of state

$$p(V - b)e^{a/RTV} = RT. \tag{11.16}$$

The equation cannot be solved in closed form for either V or T. However, using (11.14), we obtain, after some algebraic simplification:

$$\left(\frac{\partial V}{\partial T}\right)_p = -\frac{(\partial p/\partial T)_V}{(\partial p/\partial V)_T} = \left(R + \frac{a}{TV}\right) \Bigg/ \left(\frac{RT}{V - b} - \frac{a}{V^2}\right). \tag{11.17}$$

Problem 11.1.1. For the function $f(\alpha, x) = xe^{-\alpha x}$, evaluate

$$\frac{\partial f}{\partial \alpha} \quad \text{and} \quad \frac{\partial f}{\partial x}.$$

Problem 11.1.2. Show that the three partial derivatives from the Dieterici equation are consistent with the identity (11.15).

Problem 11.1.3. For the van der Waals equation of state

$$\left(p + \frac{a}{V^2}\right)(V - b) = RT.$$

evaluate $(\partial V/\partial T)_p$.

11.2 Multiple Integration

A trivial case of a double integral can be obtained from the product of two ordinary integrals, say

$$\int_a^b f(x)dx \int_c^d g(x)dx = \int_a^b f(x)dx \int_c^d g(y)dy = \int_a^b \int_c^d f(x)g(y)dx\,dy. \tag{11.18}$$

Since the variable in a definite integral is just a *dummy* variable, its name can be freely changed, say from x to y, in the first equality above. It is clearly necessary that the dummy variables have different names when they occur in a multiple integral. A double integral can also involve a nonseparable function $f(x, y)$. For well-behaved functions the integrations can be performed in either order. Thus

$$\int_{y_1}^{y_2} \int_{x_1}^{x_2} f(x, y)dx\,dy = \int_{y_1}^{y_2} \left(\int_{x_1}^{x_2} f(x, y)dx\right) dy$$

$$= \int_{x_1}^{x_2} \left(\int_{y_1}^{y_2} f(x, y)dy\right) dx. \tag{11.19}$$

For well-behaved functions the integrations above can be performed in either order.

More challenging are cases in which the limits of integration are themselves functions of x and y, for example

$$\int_{y_1}^{y_2} \left(\int_{g_1(y)}^{g_2(y)} f(x, y) dx \right) dy \quad \text{or} \quad \int_{x_1}^{x_2} \left(\int_{h_1(x)}^{h_2(x)} f(x, y) dy \right) dx. \quad (11.20)$$

If the function $f(x, y)$ is continuous, either of the integrals above can be transformed into the other by inverting the functional relations for the limits from $x = g(y)$ to $y = h(x)$. This is known as *Fubini's theorem*. The alternative evaluations of the integral are represented in Figure 11.2.

As an illustration, let us do the double integration over area involved in the geometric representation of hyperbolic functions (see Figure 4.17). Referring to Figure 11.3, it is clearly easier to first do the x-integration over horizontal strips between the straight line and the rectangular hyperbola. The area is then given by

$$A = \int_0^{y_0} \left(\int_{g_1(y)}^{g_2(y)} dx \right) dy, \quad (11.21)$$

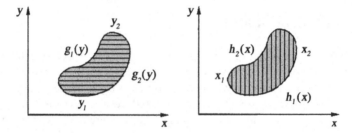

Figure 11.2 Evaluation of double integral $\int\int f(x, y)dxdy$. On left, horizontal strips are integrated over x between $g_1(y)$ and $g_2(y)$ and then summed over y. On right, vertical strips are integrated over y first. By Fubini's theorem, the alternative methods give the same result.

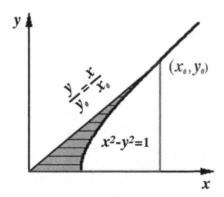

Figure 11.3 Integration over area A of the shaded crescent. This gives geometric representation of hyperbolic functions: $y_0 = \sinh(2A)$, $x_0 = \cosh(2A)$.

where

$$g_1(y) = \frac{x_0}{y_0}y \quad \text{and} \quad g_2(y) = \sqrt{1+y^2}. \tag{11.22}$$

This reduces to an integration over y:

$$A = \int_0^{y_0} \left(\sqrt{1+y^2} - \frac{x_0}{y_0}y\right) dy = \frac{1}{2}\left(y_0\sqrt{1+y_0^2} + \text{arcsinh } y_0\right) - \frac{x_0}{y_0}\frac{y_0^2}{2}. \tag{11.23}$$

Since $x_0 = \sqrt{1+y_0^2}$, we obtain $A = \frac{1}{2}\text{arcsinh } y_0$. Thus we can express the hyperbolic functions in terms of the shaded area A:

$$y_0 = \sinh(2A) \quad \text{and} \quad x_0 = \cosh(2A). \tag{11.24}$$

Problem 11.2.1. To test Fubini's theorem, redo the integration over the shaded area in Figure 11.3 using vertical, rather than horizontal strips.

11.3 Polar Coordinates

Cartesian coordinates locate a point (x, y) in a plane by specifying how far east (x-coordinate) and how far north (y-coordinate) it lies from the origin $(0,0)$. A second popular way to locate a point in two dimensions makes use of *plane polar coordinates*, (r, θ), which specifies *distance* and *direction* from the origin. As shown in Figure 11.4, the direction is defined by an angle θ, obtained by counterclockwise rotation from an eastward heading. Expressed in terms of Cartesian variables x and y, the polar coordinates are given by

$$r = \sqrt{x^2+y^2} \quad \theta = \arctan\frac{y}{x}. \tag{11.25}$$

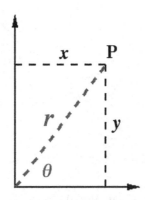

Figure 11.4 Cartesian (x, y) and polar (r, θ) coordinates of the point P.

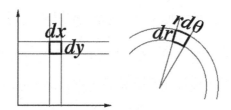

Figure 11.5 Alternative tilings of a plane in Cartesian and polar coordinates.

and conversely,

$$x = r \cos \theta \quad y = r \sin \theta. \tag{11.26}$$

Integration of a function over two-dimensional space is expressed by

$$\int_{-\infty}^{\infty} \int_{-\infty}^{\infty} f(x, y) dy \, dx. \tag{11.27}$$

In Cartesian coordinates, the plane can be "tiled" by infinitesimal rectangles of width dx and height dy. Both x and y range over $\{-\infty, \infty\}$. In polar coordinates, tiling of the plane can be accomplished by fan-shaped differential elements of area with sides dr and $r \, d\theta$, as shown in Figure 11.5. Since r and θ have ranges $\{0, \infty\}$ and $\{0, 2\pi\}$, respectively, an integral over two-dimensional space in polar coordinates is given by

$$\int_{0}^{\infty} \int_{0}^{2\pi} F(r, \theta) r \, dr \, d\theta. \tag{11.28}$$

It is understood that, expressed in terms of their alternative variables, $F(r, \theta) = f(x, y)$.

A systematic transformation of differential elements of area between two coordinate systems can be carried out using

$$dx \, dy = J \, dr \, d\theta. \tag{11.29}$$

In terms of the *Jacobian determinant*

$$J \equiv \frac{\partial(x, y)}{\partial(r, \theta)} \equiv \begin{vmatrix} (\partial x/\partial r)_\theta & (\partial y/\partial r)_\theta \\ (\partial x/\partial \theta)_r & (\partial y/\partial \theta)_r \end{vmatrix} = \begin{vmatrix} x_r & y_r \\ x_\theta & y_\theta \end{vmatrix}. \tag{11.30}$$

In this particular case, we find using (11.26)

$$J = \begin{vmatrix} \cos \theta & \sin \theta \\ -r \sin \theta & r \cos \theta \end{vmatrix} = r \cos^2 \theta + r \sin^2 \theta = r \tag{11.31}$$

in agreement with the tiling construction.

A transformation from Cartesian to polar coordinates is applied to evaluation of the famous definite integral

$$I = \int_{-\infty}^{\infty} e^{-x^2} dx.$$

Taking the square and introducing a new dummy variable, we obtain

$$I^2 = \int_{-\infty}^{\infty} e^{-x^2} dx \int_{-\infty}^{\infty} e^{-y^2} dy = \int_{-\infty}^{\infty} \int_{-\infty}^{\infty} e^{-(x^2+y^2)} dx\, dy. \tag{11.32}$$

The double integral can then be transformed to polar coordinates to give

$$I^2 = \int_0^{\infty} \int_0^{2\pi} e^{-r^2} r\, dr\, d\theta = 2\pi \int_0^{\infty} e^{-r^2} r\, dr \xrightarrow{u=r^2} 2\pi \int_0^{\infty} e^{-u} \frac{du}{2} = \pi. \tag{11.33}$$

Therefore $I = \sqrt{\pi}$.

Problem 11.3.1. Transform the Cartesian equation for a circle, $x^2 + y^2 = a^2$, into polar coordinates.

11.4 Cylindrical Coordinates

Cylindrical coordinates are a generalization of polar coordinates to three dimensions, obtained by augmenting r and θ with the Cartesian z coordinate. (Alternative notations you might encounter are r or ρ for the radial coordinate and θ or ϕ for the azimuthal coordinate.) The 3×3 Jacobian determinant is given by

$$J = \frac{\partial(x, y, z)}{\partial(r, \theta, z)} = \begin{vmatrix} x_r & y_r & z_r \\ x_\theta & y_\theta & z_\theta \\ x_z & y_z & z_z \end{vmatrix} = \begin{vmatrix} \cos\theta & \sin\theta & 0 \\ -r\sin\theta & r\cos\theta & 0 \\ 0 & 0 & 1 \end{vmatrix} = r. \tag{11.34}$$

the same value as for plane polar coordinates. An integral over three-dimensional space has the form

$$\int_0^{\infty} \int_0^{2\pi} \int_{-\infty}^{\infty} F(r, \theta, z) r\, dr\, d\theta\, dz. \tag{11.35}$$

As an application of cylindrical coordinates, let us derive the volume of a right circular cone of base radius R and altitude h, shown in Figure 11.6. This is obtained, in principle, by setting the function $F(r, \theta, z) = 1$ inside the desired volume and equal to zero everywhere else. The limits of r-integration are functions of z, such that $r(z) = Rz/h$ between $z = 0$ and $z = h$. (It is most convenient here to define the z-axis as pointing *downward* from the apex of the cone.) Thus,

$$V = \int_0^{2\pi} \left[\int_0^h \left(\int_0^{Rz/h} r\, dr \right) dz \right] d\theta = 2\pi \int_0^h \frac{R^2 z^2}{2h^2} dz = \frac{\pi R^2 h}{3}, \tag{11.36}$$

which is $\frac{1}{3}$ the volume of a cylinder with the same base and altitude.

Problem 11.4.1. Find the volume of a paraboloidal bowl of height h and rim radius R.

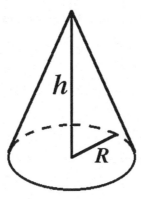

Figure 11.6 Volume of a cone. Integration in cylindrical coordinates gives $V = \frac{1}{3}\pi R^2 h$.

11.5   Spherical Polar Coordinates

Spherical polar coordinates provide the most convenient description for problems involving exact or approximate spherical symmetry. The position of an arbitrary point P is described by three coordinates (r, θ, ϕ), as shown in Figure 11.7. The radial variable r gives the distance OP from the origin to the point P. The azimuthal angle, now designated as ϕ, specifies the rotational orientation of OP about the z-axis. The third coordinate, now called θ, is the *polar angle* between OP and the Cartesian z-axis. Polar

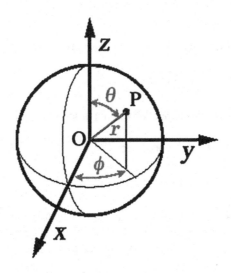

Figure 11.7 Spherical polar coordinates.

and Cartesian coordinates are connected by the relations:

$$x = r \sin\theta \cos\phi,$$
$$y = r \sin\theta \sin\phi,$$
$$z = r \cos\theta. \tag{11.37}$$

with the reciprocal relations

$$r = \sqrt{x^2 + y^2 + z^2},$$
$$\theta = \arccos\left(z/\sqrt{x^2 + y^2 + z^2}\right),$$
$$\phi = \arctan(y/x). \tag{11.38}$$

The coordinate θ is analogous to latitude in geography, in which $\theta = 0$ and $\theta = \pi$ correspond to the North and South Poles, respectively. Similarly, the angle ϕ is analogous to geographic longitude, which specifies the east or west angle with respect to the Greenwich meridian. The ranges of the spherical polar coordinates are given by:

$$0 \le r \le \infty, \qquad 0 \le \theta \le \pi, \qquad 0 \le \phi \le 2\pi.$$

The volume element in spherical polar coordinates can be determined from the Jacobian:

$$J = \frac{\partial(x, y, z)}{\partial(r, \theta, \phi)} = \begin{vmatrix} x_r & y_r & z_r \\ x_\theta & y_\theta & z_\theta \\ x_\phi & y_\phi & z_\phi \end{vmatrix} =$$

$$\begin{vmatrix} \sin\theta \cos\phi & \sin\theta \sin\phi & \cos\theta \\ r \cos\theta \cos\phi & r \cos\theta \sin\phi & -r \sin\theta \\ -r \sin\theta \sin\phi & r \sin\theta \cos\phi & 0 \end{vmatrix} = r^2 \sin\theta. \tag{11.39}$$

Therefore, a three-dimensional integral can be written

$$\int_0^\infty \int_0^\pi \int_0^{2\pi} F(r, \theta, \phi) r^2 \sin\theta \, dr \, d\theta \, d\phi. \tag{11.40}$$

A wedge-shaped differential element of volume in spherical polar coordinates is shown in Figure 11.8.

Integration over the two polar angles gives

$$\int_0^\pi \int_0^{2\pi} \sin\theta \, d\theta \, d\phi = 4\pi. \tag{11.41}$$

This represents the 4π steradians of *solid angle* which radiate from every point in three-dimensional space. For integration over a *spherical symmetrical* function $F(r)$, independent of θ and ϕ, (11.40) can be simplified to

$$\int_0^\infty F(r) 4\pi r^2 \, dr. \tag{11.42}$$

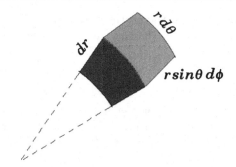

Figure 11.8 Volume element in spherical polar coordinates.

This is equivalent to integration over a series of spherical shells of area $4\pi r^2$ and thickness dr.

Problem 11.5.1. The $1s$ orbital of the hydrogen atom is described by the wavefunction $\psi(r, \theta, \phi) = (\pi a_0^3)^{-1/2} e^{-r/a_0}$, where a_0 is the Bohr radius. Calculate the integral of $|\psi|^2$ over all space.

11.6 Differential Expressions

Differential quantities of the type

$$dq(x, y) \equiv X(x, y)dx + Y(x, y)dy \tag{11.43}$$

known as *Pfaff differential expressions* are of central importance in thermodynamics. Two cases are to be distinguished. Eq. (11.43) is an *exact differential* if there exists some function $f(x, y)$ for which it is the total differential; an *inexact differential* if there exists no function which gives (11.43) upon differentiation. If dq is exact, we can write

$$d[f(x, y)] = X(x, y)dx + Y(x, y)dy. \tag{11.44}$$

Comparing with the total differential of $f(x, y)$,

$$df = \left(\frac{\partial f}{\partial x}\right)_y dx + \left(\frac{\partial f}{\partial y}\right)_x dy, \tag{11.45}$$

we can identify

$$X(x, y) = \left(\frac{\partial f}{\partial x}\right)_y \quad \text{and} \quad Y(x, y) = \left(\frac{\partial f}{\partial y}\right)_x. \tag{11.46}$$

Note further that

$$\left(\frac{\partial X}{\partial y}\right)_x = \frac{\partial^2 f}{\partial y \partial x} \quad \text{and} \quad \left(\frac{\partial Y}{\partial x}\right)_y = \frac{\partial^2 f}{\partial x \partial y}. \tag{11.47}$$

As discussed earlier, mixed second derivatives of well-behaved functions are independent of the order of differentiation. This leads to *Euler's reciprocity relation*

$$\left(\frac{\partial X}{\partial y}\right)_x = \left(\frac{\partial Y}{\partial x}\right)_y, \tag{11.48}$$

which is a necessary and sufficient condition for exactness of a differential expression. Note that the reciprocity relation neither requires nor identifies the function $f(x, y)$.

A simple example of an exact differential expression is

$$dq = y \, dx + x \, dy. \tag{11.49}$$

Here $X(x, y) = y$ and $Y(x, y) = x$, so that $(\partial X/\partial y)_x = (\partial Y/\partial x)_y = 1$ and Euler's condition is satisfied. It is easy to identify the function in this case as $f(x, y) = xy$, since $d(xy) = y \, dx + x \, dy$. A differential expression for which the reciprocity test fails is

$$dq = y \, dx - x \, dy. \tag{11.50}$$

Here $(\partial X/\partial y)_x = 1 \neq (\partial Y/\partial x)_y = -1$, so that dq is inexact and no function exists whose total differential equals (11.50).

However, an inexact differential can be cured! An inexact differential expression $X \, dx + Y \, dy$ with $\partial X/\partial y \neq \partial Y/\partial x$ can be converted into an exact differential expression by use of an *integrating factor* $g(x, y)$. This means that $g(X \, dx + Y \, dy)$ becomes exact with

$$\frac{\partial(gX)}{\partial y} = \frac{\partial(gY)}{\partial x}. \tag{11.51}$$

For example, (11.50) can be converted into an exact differential by choosing $g(x, y) = 1/y^2$ so that

$$g(x, y)dq = \frac{y \, dx - x \, dy}{y^2} = d\left(\frac{x}{y}\right). \tag{11.52}$$

Alternatively, $g(x, y) = 1/x^2$ converts the differential to $d(-y/x)$, while $g(x, y) = 1/xy$ converts it to $d \ln(x/y)$. In fact, $g(x, y)$ times any function of $f(x, y)$ is also an integrating factor. An integrating factor exists for every differential expression in two variables, such as (11.43). For differential expressions in three or more variables, such as

$$dq = \sum_{i=1}^{n} X_i \, dx_i \tag{11.53}$$

an integrating factor does *not* always exist.

The First and Second Laws of Thermodynamics can be formulated mathematically in terms of exact differentials. Individually, dq, an increment of heat gained by a system and dw, an increment of work done on a system are represented by inexact differentials. The First Law postulates that their *sum* is an exact differential:

$$dU = dq + dw, \tag{11.54}$$

which is identified with the *internal energy* U of the system. A mathematical statement of the Second Law is that $1/T$, the reciprocal of the absolute temperature, is an integrating factor for dq in a reversible process. The exact differential

$$dS = \frac{dq_{\text{rev}}}{T} \tag{11.55}$$

defines the *entropy S* of the system. These powerful generalizations hold true no matter how many independent variables are necessary to specify the thermodynamic system.

Consider the special case of reversible processes on a single-component thermodynamic system. A differential element of work in expansion or compression is given by $dw = -p\,dV$, where p is the pressure and V, the volume. Using (11.55), the differential of heat equals $dq = T\,dS$. Therefore the First Law (11.54) reduces to

$$dU = T\,dS - p\,dV \tag{11.56}$$

sometimes known as the *fundamental equation of thermodynamics*. Remarkably, since this relation contains only functions of state, U, T, S, P, and V, it applies very generally to all thermodynamic processes—reversible *and* irreversible. The structure of this differential expression implies that the energy U is a natural function of S and V, $U = U(S, V)$, and identifies the coefficients

$$T = \left(\frac{\partial U}{\partial S}\right)_V \quad \text{and} \quad p = -\left(\frac{\partial U}{\partial V}\right)_S. \tag{11.57}$$

The independent variables in a differential expression can be changed by a *Legendre tranformation*. For example, to reexpress the fundamental equation in terms of S and p, rather than S and V, we define the *enthalpy* $H \equiv U + pV$. This satisfies the differential relation

$$dH = dU + pdV + Vdp = TdS - pdV + pdV + Vdp = TdS + Vdp, \tag{11.58}$$

which must be the differential of the function $H(S, p)$. Analogously we can define the *Helmholtz free energy* $A(T, V) \equiv U - TS$, such that

$$dA = -SdT - pdV \tag{11.59}$$

and the *Gibbs free energy* $G(T, p) \equiv H - TS = U + pV - TS$, which satisfies

$$dG = -SdT + Vdp. \tag{11.60}$$

A Legendre transformation also connects the Lagrangian and Hamiltonian functions in classical mechanics. For a particle moving in one dimension, the Lagrangian $L = T - V$ can be written

$$L(x, \dot{x}) = \frac{1}{2}m\dot{x}^2 - V(x) \tag{11.61}$$

with the differential form

$$dL = \frac{\partial L}{\partial x}dx + \frac{\partial L}{\partial \dot{x}}d\dot{x}. \tag{11.62}$$

Note that

$$\frac{\partial L}{\partial \dot{x}} = m\dot{x} = p, \tag{11.63}$$

which is recognized as the momentum of the particle. The *Hamiltonian* is defined by the Legendre transformation

$$H(x, p) \equiv p\dot{x} - L(x, \dot{x}). \tag{11.64}$$

This leads to

$$H = \frac{p^2}{2m} + V(x), \tag{11.65}$$

which represents the total energy of the system.

Problem 11.6.1. Derive the reciprocity relations for Eqs. (11.58), (11.59) and (11.60) analogous to Eq. (11.57).

Problem 11.6.2. Generalize the transformation from the Lagrangian $L(\mathbf{r}, \dot{\mathbf{r}})$ to the Hamiltonian $H(\mathbf{r}, \mathbf{p})$ in three dimensions. The Lagrangian is given by

$$L(\mathbf{r}, \dot{\mathbf{r}}) = \frac{1}{2}m\mathbf{v}^2 + V(\mathbf{r}),$$

where $\mathbf{r} = \{x, y, z\}$, and $\dot{\mathbf{r}} = \mathbf{v}$, the velocity vector.

11.7 Line Integrals

The extension of the concept of integration considered in this section involves continuous summation of a differential expression along a specified path C. For the case of two independent variables, the line integral can be defined as follows:

$$\Delta q_c \equiv \int_C [X(x, y)dx + Y(x, y)dy] \equiv$$

$$\lim_{\substack{\Delta x_i, \Delta y_i \to 0 \\ n \to \infty}} \sum_{i=1}^{n} [X(x_i, y_i)\Delta x_i + Y(x_i, y_i)\Delta y_i], \tag{11.66}$$

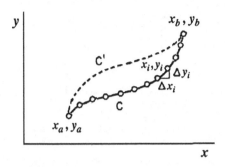

Figure 11.9 Line integral as limit of summation at points (x_i, y_i) along path C between (x_a, y_a) and (x_b, y_b). The value of the integral along path C' will, in general, be different.

where $\Delta x_i = x_i - x_{i-1}$ and $\Delta y_i = y_i - y_{i-1}$. All the points (x_i, y_i) lie on a continuous curve C connecting (x_a, y_a) to (x_b, y_b), as shown in Figure 11.9. In mechanics, the work done on a particle is equal to the line integral of the applied force along the particle's trajectory.

The line integral (11.66) reduces to a Riemann integral when the path of integration is parallel to either coordinate axis. For example along the linear path $y = y_0 = $ const, we obtain

$$\Delta q_c = \int_{x_a}^{x_b} X(x, y_0)\, dx. \tag{11.67}$$

More generally, when the curve C can be represented by a functional relation $y = g(x)$, y can be eliminated from (11.66) to give

$$\Delta q_c = \int_{x_a}^{x_b} \left\{ X[x, g(x)] + Y[x, g(x)]\frac{dg}{dx} \right\} dx. \tag{11.68}$$

In general, the value of a line integral depends on the path of integration. Thus the integrals along paths C and C' in Figure 11.9 can give different results. However, for the special case of a line integral over an exact differential, *the line integral is independent of path*, its value being determined by the initial and final points. To prove this, suppose that $X(x, y)dx + Y(x, y)dy$ is an exact differential equal to the total differential of the function $f(x, y)$. We can therefore substitute in Eq. (11.66)

$$X(x_i, y_i) \approx \frac{f(x_i, y_i) - f(x_{i-1}, y_i)}{\Delta x_i} \tag{11.69}$$

and

$$Y(x_i, y_i) \approx Y(x_{i-1}, y_i) \approx \frac{f(x_{i-1}, y_i) - f(x_{i-1}, y_{i-1})}{\Delta y_i} \tag{11.70}$$

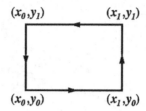

Figure 11.10 Rectangular path for integration of $dq_{\pm} = y\,dx \pm x\,dy$.

neglecting terms of higher order in Δx_i and Δy_i. Accordingly,

$$\Delta q_c \approx \sum_{i=1}^{n} \left[f(x_i, y_i) - f(x_{i-1}, y_{i-1}) \right] =$$

$$f(x_1, y_1) - f(x_0, y_0) + f(x_2, y_2) - f(x_1, y_1) + f(x_3, y_3) - f(x_2, y_2)$$

$$+ \cdots + f(x_n, y_n) - f(x_{n-1}, y_{n-1}) = f(x_n, y_n) - f(x_0, y_0).$$

$$(11.71)$$

noting that all the shaded intermediate terms cancel out. In the limit $n \to \infty$, we find therefore

$$\Delta q_c = f(x_b, y_b) - f(x_a, y_a) \tag{11.72}$$

independent of the path C, just like a Riemann integral.

Of particular significance are line integrals around closed paths, in which the initial and final points coincide. For such cyclic paths, the integral sign is written $\oint$. The closed curve is by convention traversed in the counterclockwise direction. If $dq(x, y)$ is an exact differential, then

$$\oint_C dq(x, y) = \oint_C d[f(x, y)] = f(x_0, y_0) - f(x_0, y_0) = 0 \tag{11.73}$$

for an arbitrary closed path C. (There is the additional requirement that $f(x, y)$ must be analytic within the path C, with no singularities.) When $dq(x, y)$ is inexact, the cyclic integral is, in general, different from zero. As an example, consider the integral around the rectagular closed path shown in Figure 11.10. For the two prototype examples of differential expressions $dq_{\pm} = y\,dx \pm x\,dy$ [Eqs. (11.49) and (11.50)] we find

$$\oint (y\,dx + x\,dy) = 0 \quad \text{and} \quad \oint (y\,dx - x\,dy) = 2(x_1 - x_0)(y_1 - y_0). \tag{11.74}$$

11.8   Green's Theorem

A line integral around a curve C can be related to a double Riemann integral over the enclosed area S by *Green's theorem*:

$$\oint_C [X(x, y)dx + Y(x, y)dy] = \int \int_S \left[\left(\frac{\partial Y}{\partial x}\right)_y - \left(\frac{\partial X}{\partial y}\right)_x \right] dx\, dy. \quad (11.75)$$

Green's theorem can be most easily proved by following the successive steps shown in Figure 11.11. The line integral around the small rectangle in (i) gives four contributions, which can be written

$$\oint_\square [X(x, y)dx + Y(x, y)dy] = \int_{y_0}^{y_1} [Y(x_1, y) - Y(x_0, y)]\, dy$$

$$- \int_{x_0}^{x_1} [X(x, y_1) - X(x, y_0)]\, dx. \quad (11.76)$$

However,

$$Y(x_1, y) - Y(x_0, y) = \int_{x_0}^{x_1} \frac{\partial Y}{\partial x}\, dx \quad (11.77)$$

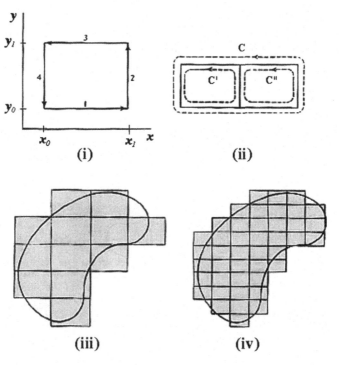

Figure 11.11 Steps in proof of Green's theorem.

and

$$X(x_1, y) - X(x_0, y) = \int_{x_0}^{x_1} \frac{\partial X}{\partial y} \, dy, \tag{11.78}$$

which establishes Green's theorem for the rectangle:

$$\oint_{\square} [X(x, y) \, dx + Y(x, y) \, dy] = \int \int_{\square} \left[\frac{\partial Y}{\partial x} - \frac{\partial X}{\partial y} \right] dx \, dy. \tag{11.79}$$

This is applicable as well for the composite figure formed by two adjacent rectangles, as in (ii). Since the line integral is taken counterclockwise in both rectangles, the common side is transversed in opposite directions along paths C and C', and the two contributions cancel. More rectangles can be added to build up the shaded figure in (iii). Green's theorem remains valid when S corresponds to the shaded area and C to its zigzag perimeter. In the final step (iv), the elements of the rectangular grid are shrunken to infinitesimal size to approach the area and perimeter of an arbitrary curved figure.

By virtue of Green's theorem, the interrelationship between differential expressions and their line integrals can be succinctly summarized. For a differential expression $dq(x, y) = X(x, y)dx + Y(x, y)dy$, any of the following statements implies the validity of the other two:

1. There exists a function $f(x, y)$ whose total differential equals $dq(x, y)$ (dq is an exact differential).
2. $\left(\frac{\partial Y}{\partial x}\right)_y = \left(\frac{\partial X}{\partial y}\right)_x$ (*Euler's reciprocity relation*).
3. $\oint_C dq(x, y) = 0$ around an arbitrary closed path C.

Problem 11.8.1.   Show that the area bounded by a smooth closed curve is given by

$$A = \frac{1}{2} \oint (y \, dx - x \, dy).$$

12 Vector Analysis

The *perceptible* physical world is three dimensional (although additional *hidden dimensions* have been speculated in superstring theories and the like). The most general mathematical representations of physical laws should therefore be relations involving three dimensions. Such equations can be compactly expressed in terms of *vectors*. Vector analysis is particularly applicable in formulating the laws of electromagnetic theory.

12.1 Scalars and Vectors

A *scalar* is a quantity which is completely described by its *magnitude*—a numerical value and usually a unit. Mass and temperature are scalars, with values, for example, like 10 kg and 300 K. A *vector* has, in addition, a *direction*. Velocity and force are vector quantities. A vector is usually printed as a boldface symbol, like **A**, while a scalar is printed in normal weight, usually in italics, like *a*. (Vectors are commonly handwritten by placing an arrow over the symbol, like $\vec{A}$ or $\vec{v}$.) A vector in three-dimensional space can be considered as a sum of three *components*. Figure 12.1 shows a vector **A** with its Cartesian components A_x, A_y, A_z, alternatively written A_1, A_2, A_3. The vector **A** is represented by the sum

$$\mathbf{A} = \mathbf{i}A_x + \mathbf{j}A_y + \mathbf{k}A_z, \tag{12.1}$$

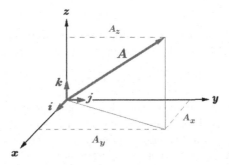

Figure 12.1 Vector **A** with Cartesian components A_x, A_y, A_z. Unit vectors **i**, **j**, **k** are also shown. The length or magnitude of **A** is given by $A = \sqrt{A_x^2 + A_y^2 + A_z^2}$.

Guide to Essential Math 2e. http://dx.doi.org/10.1016/B978-0-12-407163-6.00012-6

where $\mathbf{i}, \mathbf{j}, \mathbf{k}$ are *unit vectors* in the x, y, z directions, respectively. These are alternatively written $\hat{\mathbf{x}}, \hat{\mathbf{y}}, \hat{\mathbf{z}}$ or $\hat{\mathbf{e}}_1, \hat{\mathbf{e}}_2, \hat{\mathbf{e}}_3$. The unit vectors have magnitude one and are directed along the positive x, y, z axes, respectively. They are pure numbers, so that the units of $\mathbf{A}$ are carried by the components A_x, A_y, A_z. A vector can also be represented in matrix notation by

$$\mathbf{A} = \left[A_x, A_y, A_z \right]. \tag{12.2}$$

The magnitude of a vector $\mathbf{A}$ is written $|\mathbf{A}|$ or A. By Pythagoras' theorem in three dimensions we have

$$A = |\mathbf{A}| = \sqrt{A_x^2 + A_y^2 + A_z^2}. \tag{12.3}$$

Newton's second law, written as a vector equation

$$\mathbf{F} = m\mathbf{a}, \tag{12.4}$$

is shorthand for the *three* component relations

$$F_x = ma_x, \quad F_y = ma_y, \quad \text{and} \quad F_z = ma_z. \tag{12.5}$$

Also (12.4) implies the corresponding relation between the scalar magnitudes

$$F = ma. \tag{12.6}$$

A significant mathematical property of vector relationships is their *invariance* under translation and rotation. For example, if $\mathbf{F}$ and $\mathbf{a}$ are transformed to $\mathbf{F}'$ and $\mathbf{a}'$ by translation and/or rotation, the analog of (12.4), namely

$$\mathbf{F}' = m\mathbf{a}', \tag{12.7}$$

along with all the corresponding component relationships, is also valid. Note that scalar quantities do not change under such transformations, so $m' = m$.

A vector can be multiplied by a scalar, such that

$$c\mathbf{A} = \mathbf{i}cA_x + \mathbf{j}cA_y + \mathbf{k}cA_z. \tag{12.8}$$

For $c > 0$, this changes the magnitude of the vector while preserving its direction. For $c < 0$, the direction of the vector is reversed. Two vectors can be added using

$$\mathbf{A} + \mathbf{B} = \mathbf{i}(A_x + B_x) + \mathbf{j}(A_y + B_y) + \mathbf{k}(A_z + B_z). \tag{12.9}$$

As shown in Figure 12.2, a vector sum can be obtained either by a parallelogram construction or by a triangle construction—placing the two vectors *head to tail*. Vectors can be moved around at will, so long as their magnitudes and directions are maintained. This is called *parallel transport*. (Parallel transport is easy in Euclidean space but more complicated in curved spaces such as the surface of a sphere. Here the vector must be moved in such a way that it maintains its orientation along geodesics. Parallel transport around a closed path in a non-Euclidean space usually changes the direction of a vector.)

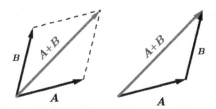

Figure 12.2 Vector sum: Parallelogram and triangle constructions.

The *position* or *displacement* vector **r** represents the three Cartesian coordinates of a point:

$$\mathbf{r} = \mathbf{i}x + \mathbf{j}y + \mathbf{k}z \quad \text{or} \quad \mathbf{r} = [x, y, z]. \tag{12.10}$$

A single symbol **r** can thus stand for the three coordinates x, y, z in the same sense that a complex number z represents the pair of numbers $\mathfrak{R}z$ and $\mathfrak{I}z$. A unit vector in the direction of **r** can be written

$$\hat{\mathbf{r}} = \frac{\mathbf{r}}{|\mathbf{r}|} = \frac{\mathbf{r}}{\sqrt{x^2 + y^2 + z^2}}. \tag{12.11}$$

A differential element of displacement can likewise be defined by

$$d\mathbf{r} = \mathbf{i}\,dx + \mathbf{j}\,dy + \mathbf{k}\,dz. \tag{12.12}$$

The notation $d\mathbf{s}$ is often used for a differential element of a curve in two- or three-dimensional space.

A function of x, y, and z can be compactly written

$$\phi(x, y, z) = \phi(\mathbf{r}). \tag{12.13}$$

If ϕ is a scalar, this represents a *scalar field*. If there is, in addition, dependence on another variable, such as time, we could write $\phi(\mathbf{r}, t)$. If the three components of a vector **A** are functions of **r**, we have a *vector field*

$$\mathbf{A}(\mathbf{r}) = \mathbf{i}A_1(\mathbf{r}) + \mathbf{j}A_2(\mathbf{r}) + \mathbf{k}A_3(\mathbf{r}) \tag{12.14}$$

or **A(r,t)** if it is time dependent.

12.2 Scalar or Dot Product

Vectors can be multipled in two different ways to give scalar products and vector products. The *scalar product*, written $\mathbf{A} \cdot \mathbf{B}$, also called the *dot product* or the *inner product*, is equal to a scalar. To see where the scalar product comes from, recall that work in mechanics equals force times displacement. If the force and displacement are not in the same direction, only the component of force along the displacement produces work. We can write $w = Fr \cos\theta$, where θ is the angle between the vectors $\mathbf{F}$ and $\mathbf{r}$.

We are thus led to define the scalar product of two vectors by

$$\mathbf{A} \cdot \mathbf{B} = AB \cos\theta. \tag{12.15}$$

When $\mathbf{B} = \mathbf{A}$, $\theta = 0$ and $\cos\theta = 1$, so the dot product reduces to the square of the magnitude of $\mathbf{A}$:

$$\mathbf{A} \cdot \mathbf{A} = A^2. \tag{12.16}$$

When $\mathbf{A}$ and $\mathbf{B}$ are orthogonal (perpendicular), $\theta = \pi/2$ and $\cos\theta = 0$, so that

$$\mathbf{A} \cdot \mathbf{B} = 0 \quad \text{when } \mathbf{A} \perp \mathbf{B}. \tag{12.17}$$

The scalar products involving the three unit vectors are therefore given by

$$\mathbf{i} \cdot \mathbf{i} = \mathbf{j} \cdot \mathbf{j} = \mathbf{k} \cdot \mathbf{k} = 1, \quad \mathbf{i} \cdot \mathbf{j} = \mathbf{j} \cdot \mathbf{k} = \mathbf{k} \cdot \mathbf{i} = 0. \tag{12.18}$$

These vectors thus constitute an *orthonormal set* with respect to scalar multiplication. To express the scalar product $\mathbf{A} \cdot \mathbf{B}$ in terms of the components of $\mathbf{A}$ and $\mathbf{B}$, write

$$\begin{aligned} \mathbf{A} \cdot \mathbf{B} &= (\mathbf{i}A_x + \mathbf{j}A_y + \mathbf{k}A_z) \cdot (\mathbf{i}B_x + \mathbf{j}B_y + \mathbf{k}B_z) \\ &= \mathbf{i} \cdot \mathbf{i} A_x B_x + \mathbf{j} \cdot \mathbf{j} A_y B_y + \mathbf{k} \cdot \mathbf{k} A_z B_z + \mathbf{i} \cdot \mathbf{j} A_x B_y + \cdots . \end{aligned} \tag{12.19}$$

Using (12.18) for the products of unit vectors, we find

$$\mathbf{A} \cdot \mathbf{B} = A_x B_x + A_y B_y + A_z B_z = \sum_{i=1}^{3} A_i B_i. \tag{12.20}$$

The scalar product has the same structure as the three-dimensional matrix product of a row vector with a column vector:

$$\widetilde{\mathbb{A}}\mathbb{B} = A_1 B_1 + A_2 B_2 + A_3 B_3. \tag{12.21}$$

12.3   Vector or Cross Product

Consider a point P rotating counterclockwise about a vertical axis through the origin, as shown in Figure 12.3. Let $\mathbf{r}$ be the vector from the origin to point P and $\mathbf{v}$, the instantaneous linear velocity of the point as it moves around a circle of radius $r \sin\theta$. A rotation is conventionally represented by an *axial vector* $\boldsymbol{\omega}$ normal to the plane of motion, such that the trajectory of P winds about $\boldsymbol{\omega}$ in a counterclockwise sense, the same as the direction a right-handed screw advances as it is turned. This is remembered most easily by using the *right-hand rule* shown in Figure 12.4. A rotational velocity of ω radians/s moves point P with a speed $v = \omega r \sin\theta$ around the circle. The velocity

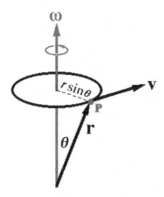

Figure 12.3 Rotation of point P about vertical axis, represented by the vector product relation $\mathbf{v} = \boldsymbol{\omega} \times \mathbf{r}$.

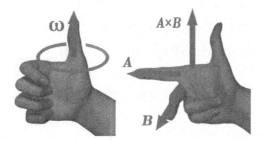

Figure 12.4 Right-hand rules. Left: Direction of axial vector $\boldsymbol{\omega}$ representing counterclockwise rotation. Right: Direction of vector product $\mathbf{A} \times \mathbf{B}$.

vector $\mathbf{v}$ thus has magnitude v and instantaneous direction normal to both $\mathbf{r}$ and $\boldsymbol{\omega}$. This motivates definition of the *vector product*, also known as the *cross product* or the *outer product*, such that

$$\mathbf{v} = \boldsymbol{\omega} \times r, \quad |\mathbf{v}| = v = \omega r \sin \theta. \tag{12.22}$$

The direction of $\mathbf{v}$ is determined by the counterclockwise rotation of the first vector $\boldsymbol{\omega}$ into the second $\mathbf{r}$, shown also in Figure 12.4.

As another way to arrive at the vector product, consider the parallelogram with adjacent sides formed by vectors $\mathbf{A}$ and $\mathbf{B}$, as shown in Figure 12.5. The area of the parallelogram is equal to its base A times its altitude $B \sin \theta$. By definition, the vector product $\mathbf{A} \times \mathbf{B}$ has magnitude $AB \sin \theta$ and direction *normal* to the parallelogram. The operation of vector multiplication is *anticommutative* since

$$\mathbf{B} \times \mathbf{A} = -\mathbf{A} \times \mathbf{B}. \tag{12.23}$$

This implies that the cross product of a vector with itself equals zero:

$$\mathbf{A} \times \mathbf{A} = 0. \tag{12.24}$$

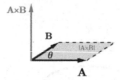

Figure 12.5 Vector product $\mathbf{A} \times \mathbf{B}$. The parallelogram has area $| \mathbf{A} \times \mathbf{B} | = AB \sin \theta$.

By contrast, scalar multiplication is *commutative* with

$$\mathbf{B} \cdot \mathbf{A} = \mathbf{A} \cdot \mathbf{B}. \tag{12.25}$$

The three unit vectors have the following vector products:

$$\mathbf{i} \times \mathbf{i} = \mathbf{j} \times \mathbf{j} = \mathbf{k} \times \mathbf{k} = 0, \quad \mathbf{i} \times \mathbf{j} = -\mathbf{j} \times \mathbf{i} = \mathbf{k},$$
$$\mathbf{j} \times \mathbf{k} = -\mathbf{k} \times \mathbf{j} = \mathbf{i}, \quad \mathbf{k} \times \mathbf{i} = -\mathbf{i} \times \mathbf{k} = \mathbf{j}. \tag{12.26}$$

Therefore, the cross product of two vectors in terms of their components can be determined from

$$\mathbf{A} \times \mathbf{B} = (\mathbf{i}A_x + \mathbf{j}A_y + \mathbf{k}A_z) \times (\mathbf{i}B_x + \mathbf{j}B_y + \mathbf{k}B_z)$$
$$= \mathbf{i}(A_y B_z - A_z B_y) + \mathbf{j}(A_z B_x - A_x B_z) + \mathbf{k}(A_x B_y - A_y B_x). \tag{12.27}$$

This can be compactly represented as a 3×3 determinant:

$$\mathbf{A} \times \mathbf{B} = \begin{vmatrix} \mathbf{i} & \mathbf{j} & \mathbf{k} \\ A_x & A_y & A_z \\ B_x & B_y & B_z \end{vmatrix}. \tag{12.28}$$

The individual components are given by

$$(\mathbf{A} \times \mathbf{B})_x = A_y B_z - A_z B_y \quad et \ cyc, \tag{12.29}$$

where "*et cyc*" stands for the other two relations obtained by cyclic permutations $x \to y \to z \to x$. Another way to write a cross product is

$$(\mathbf{A} \times \mathbf{B})_i = \sum_{j,k=1}^{3} \epsilon_{ijk} A_j B_k, \tag{12.30}$$

in terms of the *Levi-Civita symbol* ϵ_{ijk}, defined by

$$\epsilon_{123} = \epsilon_{231} = \epsilon_{312} = +1,$$
$$\epsilon_{321} = \epsilon_{213} = \epsilon_{132} = -1,$$
$$\epsilon_{ijk} = 0 \quad \text{otherwise}. \tag{12.31}$$

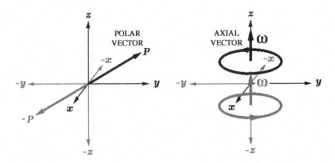

Figure 12.6 Inversion behavior of polar vector **P** and axial vector ω. Axes and vectors before and after inversion are drawn.

The vector product occurs in the force law for a charged particle moving in an electromagnetic field. A particle with charge q moving with velocity **v** in an electric field **E** and a magnetic induction field **B** experiences a *Lorentz force*:

$$\mathbf{F} = q\mathbf{E} + q\mathbf{v} \times \mathbf{B}. \tag{12.32}$$

The magnetic component of the force is perpendicular to both the velocity and the magnetic induction, which causes charged particles to deflect into curved paths. This underlies the principle of the cyclotron and other particle accelerators.

Inversion of a coordinate system means reversing the directions if the x, y, z axes, or equivalently replacing x, y, z by $-x$, $-y$, $-z$. This amounts to turning a right-handed into a left-handed coordinate system. A *polar vector* is transformed into its negative by inversion, as shown in Figure 12.6. A circulation in three dimensions, which can be represented by an *axial vector*, remains, by contrast, *unchanged* under inversion. An axial vector is also called a *pseudovector* to highlight its different inversion symmetry. The vector product of two polar vectors gives an axial vector: symbolically, $\mathbf{P} \times \mathbf{P} = \mathbf{A}$. You can show also that $\mathbf{P} \times \mathbf{A} = \mathbf{P}$ and $\mathbf{A} \times \mathbf{A} = \mathbf{A}$. In electromagnetic theory, the electric field **E** is a polar vector while the magnetic induction **B** is an axial vector. This is consistent with the fact that magnetic fields originate from circulating electric charges.

Problem 12.3.1. A charged particle in an electromagnetic field is described by the Lagrangian

$$L(\mathbf{r}, \mathbf{v}) = \frac{1}{2}mv^2 - q\mathbf{E}(\mathbf{r}) + q\mathbf{v} \cdot \mathbf{A}(\mathbf{r}),$$

where q is the electric charge and $\mathbf{B} = \nabla \times \mathbf{A}$, where **A** is the vector potential. As a generalization of the result of Problem 11.6.1, derive the corresponding Hamiltonian $H(\mathbf{r}, \mathbf{p})$.

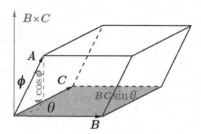

Figure 12.7 Triple scalar product as volume of parallelepiped: base area $= BC\sin\theta$, altitude $= A\cos\phi$, volume $= ABC\sin\theta\cos\phi = |\mathbf{A} \cdot (\mathbf{B} \times \mathbf{C})|$.

12.4 Triple Products of Vectors

The *triple scalar product*, given by

$$\mathbf{A} \cdot (\mathbf{B} \times \mathbf{C}) = (\mathbf{A} \times \mathbf{B}) \cdot \mathbf{C} = \begin{vmatrix} A_x & A_y & A_z \\ B_x & B_y & B_z \\ C_x & C_y & C_z \end{vmatrix}, \tag{12.33}$$

represents the volume of a parallelepiped formed by the three vectors $\mathbf{A}$, $\mathbf{B}$, and $\mathbf{C}$, as shown in Figure 12.7. A 3×3 determinant changes sign when two rows are interchanged but preserves its value under a cyclic permutation, so that

$$\mathbf{A}\cdot(\mathbf{B}\times\mathbf{C}) = \mathbf{B}\cdot(\mathbf{C}\times\mathbf{A}) = \mathbf{C}\cdot(\mathbf{A}\times\mathbf{B}) = -\mathbf{C}\cdot(\mathbf{B}\times\mathbf{A}) = -\mathbf{B}\cdot(\mathbf{A}\times\mathbf{C}) = -\mathbf{A}\cdot(\mathbf{C}\times\mathbf{B}). \tag{12.34}$$

For three polar vectors, the triple scalar product changes sign upon inversion. Such a quantity is known as a *pseudoscalar*, in contrast to a scalar, which is invariant to inversion.

You might also encounter the *triple vector product* $\mathbf{A} \times (\mathbf{B} \times \mathbf{C})$, which is a *vector* quantity. This can be evaluated using the Levi-Civita representation (12.30). The i component of the triple product can be written

$$[\mathbf{A} \times (\mathbf{B} \times \mathbf{C})]_i = \sum_{jk} \epsilon_{ijk} A_j (\mathbf{B} \times \mathbf{C})_k = \sum_{jk\ell m} \epsilon_{ijk}\epsilon_{k\ell m} A_j B_\ell C_m, \tag{12.35}$$

where we have introduced new dummy indices as needed. Focus on the sum over k of the product of the Levi-Civita symbols $\sum_k \epsilon_{ijk}\epsilon_{k\ell m}$. For nonzero contributions, i and j must be different from k, and likewise for ℓ and m. We must have either that $i = \ell, j = m$ or $i = m, j = \ell$. Therefore

$$\sum_k \epsilon_{ijk}\epsilon_{k\ell m} = \delta_{i\ell}\delta_{jm} - \delta_{im}\delta_{j\ell}. \tag{12.36}$$

Since a sum containing a Kronecker delta reduces to a single term, we find

$$[\mathbf{A} \times (\mathbf{B} \times \mathbf{C})]_i = B_i \sum_j A_j C_j - C_i \sum_m A_m B_m = B_i(\mathbf{A}\cdot\mathbf{C}) - C_i(\mathbf{A}\cdot\mathbf{B}). \tag{12.37}$$

Therefore, in full vector notation,

$$\mathbf{A} \times (\mathbf{B} \times \mathbf{C}) = \mathbf{B}(\mathbf{A} \cdot \mathbf{C}) - \mathbf{C}(\mathbf{A} \cdot \mathbf{B}), \tag{12.38}$$

which has the popular mnemonic "BAC(K) minus CAB."

12.5   Vector Velocity and Acceleration

A particle moving in three dimensions can be represented by a time-dependent displacement vector:

$$\mathbf{r}(t) = \mathbf{i}x(t) + \mathbf{j}y(t) + \mathbf{k}z(t) \quad \text{or} \quad \mathbf{r}(t) = [x(t), y(t), z(t)]. \tag{12.39}$$

The *velocity* of the particle is the time derivative of $\mathbf{r}(t)$

$$\mathbf{v}(t) = \frac{d\mathbf{r}(t)}{dt} = \dot{\mathbf{r}}(t) = \lim_{\Delta t \to 0} \left[\frac{\mathbf{r}(t + \Delta t) - \mathbf{r}(t)}{\Delta t} \right], \tag{12.40}$$

as represented in Figure 12.8. The Cartesian components of velocity are

$$v_x = \frac{dx}{dt} = \dot{x}(t), \quad v_y = \frac{dy}{dt} = \dot{y}(t), \quad v_z = \frac{dz}{dt} = \dot{z}(t). \tag{12.41}$$

The magnitude of the velocity vector is the *speed*:

$$v = |\mathbf{v}| = \sqrt{v_x^2 + v_y^2 + v_z^2}. \tag{12.42}$$

The velocity vector will be parallel to the displacement only if the particle is moving in a straight line. In Figure 12.8, $\Delta \mathbf{r} = \mathbf{r}(t + \Delta t) - \mathbf{r}(t)$ can clearly be in a different direction. Thus, $\mathbf{v}(t)$ need not, in general, be parallel to $\mathbf{r}(t)$.

Acceleration is the time derivative of velocity. We find

$$\mathbf{a}(t) = \frac{d\mathbf{v}}{dt} = \dot{\mathbf{v}}(t) = \ddot{\mathbf{r}}(t) = \lim_{\Delta t \to 0} \left[\frac{\mathbf{v}(t + \Delta t) - \mathbf{v}(t)}{\Delta t} \right] \tag{12.43}$$

with Cartesian components

$$a_x = \frac{dv_x}{dt} = \frac{d^2x}{dt^2} = \ddot{x}(t), \quad \text{etc.} \tag{12.44}$$

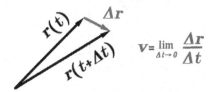

Figure 12.8 Velocity vector.

12.6   Circular Motion

Figure 12.9 represents a particle moving in a circle of radius r in the x, y-plane. Assuming a constant angular velocity of ω radians/s, the angular position of the particle is given by $\theta = \omega t$. Cartesian coordinates can be defined with $x(t) = r\cos\omega t$, $y(t) = r\cos\omega t$, $z(t) = 0$, so that

$$\mathbf{r}(t) = \mathbf{i}r\cos\omega t + \mathbf{j}r\cos\omega t. \tag{12.45}$$

Differentiation with respect to t gives the velocity

$$\mathbf{v}(t) = -\mathbf{i}\omega r\sin\omega t + \mathbf{j}\omega r\sin\omega t = -\mathbf{i}\omega y(t) + \mathbf{j}\omega x(t). \tag{12.46}$$

As we have seen, the axial vector representing angular velocity is given by $\boldsymbol{\omega} = \mathbf{k}\omega$. Thus,

$$\boldsymbol{\omega} \times \mathbf{r} = \mathbf{k}\omega \times (\mathbf{i}x + \mathbf{j}y) = -\mathbf{i}\omega y + \mathbf{j}\omega x \tag{12.47}$$

and

$$\mathbf{v} = \boldsymbol{\omega} \times \mathbf{r}, \tag{12.48}$$

in agreement with Eq. (12.22).

To find the acceleration of a particle in uniform circular motion, note that $\boldsymbol{\omega}$ is a time-independent vector, so that

$$\frac{d\mathbf{v}}{dt} = \boldsymbol{\omega} \times \frac{d\mathbf{r}}{dt} \;\Rightarrow\; \mathbf{a} = \boldsymbol{\omega} \times \mathbf{v}. \tag{12.49}$$

In this case acceleration represents a change in the *direction* of the velocity vector, while the magnitude v remains constant. Using the right-hand rule, the acceleration is seen to be directed toward the center of the circle. This is known as *centripetal acceleration*, centripetal meaning "center seeking." The magnitude of the centripetal acceleration is

$$a = \omega v = \omega^2 r = \frac{v^2}{r}. \tag{12.50}$$

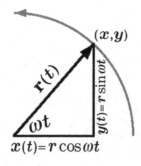

$$x(t) = r\cos\omega t$$

Figure 12.9 Displacement vector $\mathbf{r}(t)$ for particle moving in a circular path with angular velocity ω. The x- and y-components of $\mathbf{r}$ are shown.

The force which causes centripetal acceleration is called *centripetal force*. By Newton's second law

$$F_{\text{centripetal}} = \frac{mv^2}{r}. \tag{12.51}$$

For example, gravitational attraction to the Sun provides the centripetal force which keeps planets in nearly elliptical orbits. The effect known as *centrifugal force* is actually an artifice. It actually represents the tendency of a body to continue moving in a straight line, according to Newton's first law. It might be naïvely perceived as a force *away* from the center. Actually, it is the *centripetal* force which keeps a body moving along a curved path, in resistance to this inertial tendency.

12.7 Angular Momentum

The *linear momentum* $\mathbf{p} = m\mathbf{v} = m\dot{\mathbf{r}}$ is a measure of inertia for a particle moving in a straight line. Accordingly, Newton's second law can be elegantly written

$$\mathbf{F} = \frac{d\mathbf{p}}{dt}. \tag{12.52}$$

A result applying to angular motion can be obtained by taking the vector product of Newton's law:

$$\mathbf{r} \times \mathbf{F} = \mathbf{r} \times \frac{d\mathbf{p}}{dt} = \frac{d}{dt}(\mathbf{r} \times \mathbf{p}). \tag{12.53}$$

The last equality follows from

$$\frac{d}{dt}(\mathbf{r} \times \mathbf{p}) = \frac{d\mathbf{r}}{dt} \times \mathbf{p} + \mathbf{r} \times \frac{d\mathbf{p}}{dt} = \mathbf{v} \times m\mathbf{v} + \mathbf{r} \times \frac{d\mathbf{p}}{dt}, \tag{12.54}$$

and the fact that $\mathbf{v} \times \mathbf{v} = 0$. *Angular momentum*—more precisely, *orbital* angular momentum—is defined by

$$\mathbf{L} \equiv \mathbf{r} \times \mathbf{p}. \tag{12.55}$$

The analog of Eq. (12.52) for angular motion is

$$\boldsymbol{\tau} = \frac{d\mathbf{L}}{dt}, \tag{12.56}$$

where $\boldsymbol{\tau} = \mathbf{r} \times \mathbf{F}$ is called the *torque* or turning force. The law of *conservation of angular momentum* implies that, in the absence of external torque, a system will continue in its rotational motion.

Consider a particle of mass m in angular motion with the angular velocity vector $\boldsymbol{\omega}$. The angular momentum can be related to the angular velocity using Eqs. (12.48) and (12.38):

$$\mathbf{L} = m\mathbf{r} \times \mathbf{v} = m\mathbf{r} \times (\boldsymbol{\omega} \times \mathbf{r}) = m\omega r^2 - m\mathbf{r}(\boldsymbol{\omega} \cdot \mathbf{r}) = \mathbb{I} \cdot \boldsymbol{\omega}, \tag{12.57}$$

where $\mathbb{I}$ is the *moment of inertia tensor*, represented by the 3×3 matrix

$$\mathbb{I} = \begin{bmatrix} m(y^2 + z^2) & -mxy & -mxz \\ -myx & m(z^2 + x^2) & -myz \\ -mzx & -mzy & m(x^2 + y^2) \end{bmatrix}. \tag{12.58}$$

Tensors are the next member of the hierarchy which begins with scalars and vectors. The dot product of a tensor with a vector gives another vector. Usually the moment of inertia is defined for a *rigid body*, a system of particles with fixed relative coordinates. We must then replace $m(x^2 + y^2)$ in the matrix by the corresponding summation over all particles, $\sum_i m_i(y_i^2 + z_i^2)$, and so forth. We will focus on the simple case of circular motion, with $\mathbf{r}$ being perpendicular to $\boldsymbol{\omega}$ as in Figure 12.9. This implies that $\boldsymbol{\omega}\cdot\mathbf{r} = 0$, so that the moment of inertia reduces to a scalar:

$$\mathbf{L} = I\boldsymbol{\omega} \quad \text{with } I = mr^2. \tag{12.59}$$

Moment of inertia is a measure of a system's resistance to change in its rotational motion, in the same way that mass is a measure of resistance to change in linear motion.

The kinetic energy for a mass in circular motion can be expressed in terms of the angular frequency. We find

$$T_{\text{rot}} = \frac{1}{2}mv^2 = \frac{1}{2}m(\omega r)^2 = \frac{1}{2}I\omega^2. \tag{12.60}$$

This can also be expressed

$$T_{\text{rot}} = \frac{L^2}{2I}. \tag{12.61}$$

Evidently, relations for linear motion can be transformed to their analogs for angular motion with the substitutions: $m \to I, x \to \theta, v \to \omega, p \to L$. Here is a handy table giving the linear and angular equivalents for some mechanical variables:

Variable	Linear	Angular
Displacement	x	θ
Velocity	$v = dx/dt$	$\omega = d\theta/dt$
Acceleration	$a = d^2x/dt^2$	$\alpha = d^2\theta/dt^2$
Inertial variable	m	$I = mR^2$
Momentum variable	$p = mv$	$L = I\omega$
Force (torque)	$F = dp/dt = ma$	$\tau = dL/dt = I\alpha$
Kinetic energy	$\frac{1}{2}mv^2$ or $p^2/2m$	$\frac{1}{2}I\omega^2$ or $L^2/2I$

12.8 Gradient of a Scalar Field

A scalar field $\phi(\mathbf{r}) = \phi(x, y, z)$ can be represented graphically by a family of surfaces on which the field has a sequence of constant values (see Figure 12.10). If the

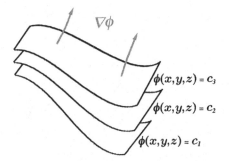

Figure 12.10 Gradients $\nabla\phi$ of a scalar field shown as red arrows. The gradient at every point x, y, z is normal to the surface of constant ϕ in the direction of maximum increase of $\phi(x, y, z)$.

scalar quantity is temperature, the surfaces are called *isotherms*. Analogously, barometric pressure is represented by *isobars* and electrical potential by *equipotentials*. (The less-familiar generic term for constant-value surfaces of a scalar field is *isopleths*.) When the position vector $\mathbf{r}$ is changed by an infinitesimal $d\mathbf{r}$, the change in $\phi(\mathbf{r})$ is given by the total differential [see Eq. (11.12)]

$$d\phi = \frac{\partial\phi}{\partial x}dx + \frac{\partial\phi}{\partial y}dy + \frac{\partial\phi}{\partial z}dz. \tag{12.62}$$

This has the form of a scalar product of the differential displacement $d\mathbf{r} = \mathbf{i}\,dx + \mathbf{j}\,dy + \mathbf{k}\,dz$ with the *gradient vector*

$$\nabla\phi(x, y, z) \equiv \mathbf{i}\frac{\partial\phi}{\partial x} + \mathbf{j}\frac{\partial\phi}{\partial y} + \mathbf{k}\frac{\partial\phi}{\partial z}. \tag{12.63}$$

In particular

$$d\phi = \nabla\phi \cdot d\mathbf{r}. \tag{12.64}$$

When $d\mathbf{r}$ happens to lie within one of the surfaces of constant ϕ, then $d\phi = 0$, which implies that the gradient $\nabla\phi$ at every point is *normal* to the surface $\phi = $ const containing that point. This is shown in Figure 12.10, with arrows representing gradient vectors. Where constant ϕ surfaces are closer together, the function must be changing rapidly and the magnitude of $\nabla\phi$ is correspondingly larger. This is more easily seen in a two-dimensional contour map, Figure 12.11. The gradient at x, y, z points in the direction of *maximum* increase of $\phi(x, y, z)$. The symbol ∇ is called "del" or "nabla" and stands for the vector differential operator

$$\nabla \equiv \mathbf{i}\frac{\partial}{\partial x} + \mathbf{j}\frac{\partial}{\partial y} + \mathbf{k}\frac{\partial}{\partial z}. \tag{12.65}$$

Sometimes $\nabla\phi$ is written grad ϕ.

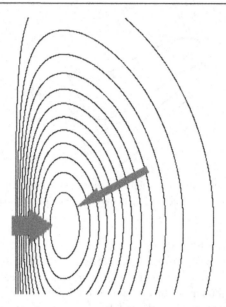

Figure 12.11 Two-dimensional contour map showing, for example, the elevation around a hill. Where the contours are more closely spaced, the terrain is steeper and the gradient larger in magnitude. This is indicated by the thicker arrow.

The change in a scalar field $\phi(\mathbf{r})$ along a unit vector $\hat{\mathbf{u}}$ is called the *directional derivative*, defined by

$$\nabla_u \phi = \nabla \phi \cdot \hat{\mathbf{u}} = \lim_{\epsilon \to 0} \frac{\phi(\mathbf{r} + \epsilon \hat{\mathbf{u}}) - \phi(\mathbf{r})}{\epsilon}. \tag{12.66}$$

A finite change in the scalar field $\phi(\mathbf{r})$ as $\mathbf{r}$ moves from $\mathbf{r}_1$ to $\mathbf{r}_2$ over a path C is given by the line integral

$$\Delta \phi = \int_C \nabla \phi \cdot d\mathbf{r}. \tag{12.67}$$

Work in mechanics is given by a line integral over force: $w = \int_C \mathbf{F} \cdot d\mathbf{r}$. The work done *by* the system is the negative of this quantity. Thus, for a conservative system, force is given by the negative gradient of potential energy: $\mathbf{F}(\mathbf{r}) = -\nabla V(\mathbf{r})$. An analogous relation holds in electrostatics between the electric field and the potential: $\mathbf{E} = -\nabla \Phi$.

Problem 12.8.1.   The force between two electrical charges q_1 and q_2 separated by a displacement vector $\mathbf{r}$ (in three dimensions) is given by Coulomb's law, $\mathbf{F} = -q_1 q_2 \mathbf{r}/r^3$. Derive the potential-energy function $V(\mathbf{r})$.

12.9   Divergence of a Vector Field

One of the cornerstones of modern physics is the *conservation of electric charge*. If an element of volume contains a certain quantity of charge, that quantity can change only when charge flows through the boundary of the volume. The charge per unit volume measured in Coulombs per unit volume (C/m^3), is designated the *charge density* $\rho(\mathbf{r}, t)$. The flow of charge across unit area per unit time is represented by the *current density* $\mathbf{J}(\mathbf{r}, t)$. The current density at a point $\mathbf{r}$ equals the product of the density and the instantaneous velocity of charge at that point:

$$\mathbf{J}(\mathbf{r}, t) = \rho(\mathbf{r}, t)\mathbf{v}(\mathbf{r}, t). \tag{12.68}$$

This is dimensionally consistent since

$$\frac{C}{m^2\,s} = \frac{C}{m^3} \times \frac{m}{s}. \tag{12.69}$$

Note that $\mathbf{J}(\mathbf{r}, t)$ and $\mathbf{v}(\mathbf{r}, t)$ are vector fields, while $\rho(\mathbf{r}, t)$ is a scalar field.

Consider the element of volume $\Delta V = \Delta x \Delta y \Delta z$ in Cartesian coordinates, shown in Figure 12.12. Designating the coordinates at the center by (x, y, z), the middle of the faces normal to the x-direction are $(x+\frac{1}{2}\Delta x, y, z)$ and $(x-\frac{1}{2}\Delta x, y, z)$, and analogously for the faces normal to the other two directions. The electric charge contained in ΔV can be approximated by $\rho(x, y, z, t)\Delta V$. The net charge leaving ΔV per unit time is then equal to the integral over the current density normal to the surface ΔS enclosing the element of volume. Such a surface integral can be written $\int_{\Delta S} \mathbf{J}\cdot d\boldsymbol{\sigma}$, where $d\boldsymbol{\sigma} = \hat{\mathbf{n}}\,d\sigma$, an element of area with the direction of the outward normal from the surface. In the present case, the surface integral is the sum of contributions from the six faces of ΔV, thus

$$\int_{\Delta S} \mathbf{J}\cdot d\boldsymbol{\sigma} \approx \left[J_x(x+\tfrac{1}{2}\Delta x, y, z) - J_x(x-\tfrac{1}{2}\Delta x, y, z) \right] \Delta y \Delta z +$$
$$\left[J_y(x, y+\tfrac{1}{2}\Delta y, z) - J_y(x, y-\tfrac{1}{2}\Delta y, z) \right] \Delta z \Delta x +$$
$$\left[J_z(x, y, z+\tfrac{1}{2}\Delta z) - J_z(x, y, z-\tfrac{1}{2}\Delta z) \right] \Delta x \Delta y. \tag{12.70}$$

Divide each side by ΔV and let the element of volume be shrunken to a point. In the limit $\Delta x, \Delta y, \Delta z \to 0$, the terms containing J_x reduce to a partial derivative, as follows:

$$\frac{J_x(x+\tfrac{1}{2}\Delta x, y, z) - J_x(x-\tfrac{1}{2}\Delta x, y, z)}{\Delta x} \xrightarrow{\Delta x \to 0} \frac{\partial J_x}{\partial x}. \tag{12.71}$$

Adding the analogous y and z contributions, we find

$$\lim_{\Delta V \to 0} \frac{\int_{\Delta S} \mathbf{J}\cdot d\boldsymbol{\sigma}}{\Delta V} = \frac{\partial J_x}{\partial x} + \frac{\partial J_y}{\partial y} + \frac{\partial J_z}{\partial z}. \tag{12.72}$$

The right-hand side has the structure of a scalar product of ∇ with $\mathbf{J}$:

$$\nabla\cdot\mathbf{J} = \left(\mathbf{i}\frac{\partial}{\partial x} + \mathbf{j}\frac{\partial}{\partial y} + \mathbf{k}\frac{\partial}{\partial z} \right)\cdot\left(\mathbf{i}J_x + \mathbf{j}J_y + \mathbf{k}J_z \right) = \frac{\partial J_x}{\partial x} + \frac{\partial J_y}{\partial y} + \frac{\partial J_z}{\partial z}. \tag{12.73}$$

called the *divergence* of $\mathbf{J}$ (also written div $\mathbf{J}$).

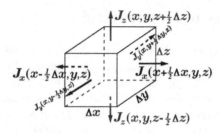

Figure 12.12 Flux of a vector field **J** out of an element of volume ΔV with surface ΔS, for derivation of the divergence theorem.

The divergence of the current density at a point x,y,z represents the *net outward flux* of electric charge from that point. Since electric charge is conserved, the flow of charge from every point must be balanced by a reduction of the charge density $\rho(\mathbf{r}, t)$ in the vicinity of that point. This leads to the *equation of continuity*

$$\frac{\partial \rho}{\partial t} + \nabla \cdot \mathbf{J} = 0. \tag{12.74}$$

In the *steady-state* case, in which there is no accumulation of charge at any point, the equation of continuity reduces to

$$\nabla \cdot \mathbf{J} = 0 \quad \text{(Steady state)}. \tag{12.75}$$

In defining the divergence of a vector field $\mathbf{A}(\mathbf{r}, t)$, we have transformed an integration over surface area into an integration over volume, as shown in Figure 12.12. When two such elements of volume are adjacent, the contribution from their interface cancels out, since flux into one element is exactly canceled by flux out of the adjacent element. By adding together an infinite number of infinitesimal elements, the result can be generalized to a finite volume of arbitrary shape. We arrive thereby at the *divergence theorem* (also known as Gauss' theorem):

$$\int_S \mathbf{A} \cdot d\boldsymbol{\sigma} = \int_V \nabla \cdot \mathbf{A} \, d\tau. \tag{12.76}$$

A differential element of volume is conventionally written $d\tau$. Sometimes the notation ∂V is used for S, to indicate the surface area enclosing the volume V.

The divergence of the gradient of a scalar field occurs in several fundamental equations of electromagnetism, wave theory, and quantum mechanics. In Cartesian coordinates

$$\nabla \cdot \nabla \phi(\mathbf{r}) = \frac{\partial^2 \phi}{\partial x^2} + \frac{\partial^2 \phi}{\partial y^2} + \frac{\partial^2 \phi}{\partial z^2}. \tag{12.77}$$

The operator

$$\nabla \cdot \nabla \equiv \nabla^2 = \frac{\partial^2}{\partial x^2} + \frac{\partial^2}{\partial y^2} + \frac{\partial^2}{\partial z^2} \tag{12.78}$$

is called the *Laplacian* or "del squared." Sometimes ∇^2 is abbreviated as Δ.

12.10   Curl of a Vector Field

As we have seen, a vector field can be imagined as some type of *flow*. We used this idea in defining the divergence. Another aspect of flow might be *circulation*, in the sense of a local angular velocity. Figure 12.13 gives a schematic pictorial representation of vector fields with outward flux (left) and with circulation (right). To be specific, consider a rectangular element in the x, y-plane as shown in Figure 12.14. The center of the rectangle at (x, y), the four sides are at $x \pm \frac{1}{2}\Delta x$ and $y \pm \frac{1}{2}\Delta y$. The area if the rectangle is $\Delta \sigma = \Delta x, \Delta y$. The circulation of a vector field $\mathbf{A}(\mathbf{r})$ around the rectangle in the counterclockwise sense is approximated by

$$\oint_{\Delta\sigma} \mathbf{A}(\mathbf{r}) \cdot ds \approx A_x \left(x, y - \tfrac{1}{2}\Delta y \right) \Delta x + A_y \left(x + \tfrac{1}{2}\Delta x, y \right) \Delta y$$
$$-A_x \left(x, y + \tfrac{1}{2}\Delta y \right) \Delta x - A_y \left(x - \tfrac{1}{2}\Delta x, y \right) \Delta y. \qquad (12.79)$$

Next we divide both sides by $\Delta \sigma = \Delta x \Delta y$ and take the limits $\Delta x, \Delta y \to 0$. Since

$$\lim_{\Delta x \to 0} \left[\frac{A_y \left(x + \tfrac{1}{2}\Delta x, y \right) - A_y \left(x - \tfrac{1}{2}\Delta x, y \right)}{\Delta x} \right] = \frac{\partial A_y}{\partial x} \qquad (12.80)$$

and

$$\lim_{\Delta y \to 0} \left[\frac{A_x(x, y + \tfrac{1}{2}\Delta y) - A_x(x, y - \tfrac{1}{2}\Delta y)}{\Delta y} \right] = \frac{\partial A_x}{\partial y}, \qquad (12.81)$$

we find

$$\lim_{\Delta\sigma \to 0} \frac{\oint_{\Delta a} \mathbf{A}(\mathbf{r}) \cdot ds}{\Delta \sigma} = \frac{\partial A_y}{\partial x} - \frac{\partial A_x}{\partial y} = (\nabla \times \mathbf{A})_z. \qquad (12.82)$$

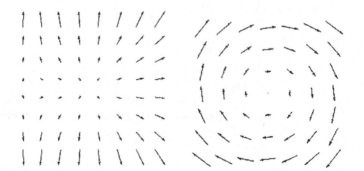

Figure 12.13 Schematic representations of vector fields with divergence (left) and curl (right).

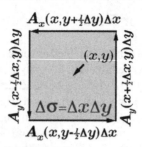

Figure 12.14 Circulation of a vector field $\mathbf{A(r)}$ about an element of area $\Delta\sigma$. In the limit $\Delta\sigma \to 0$, this gives the z-component of the curl $\nabla \times \mathbf{A}$.

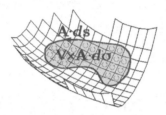

Figure 12.15 Stokes' theorem: $\oint_C A \cdot ds = \int_S (\nabla \times \mathbf{A}) \cdot d\boldsymbol{\sigma}$.

It can be recognized that the difference of partial derivatives represents the z-component of a cross product involving the vector operator ∇:

$$
\nabla \times \mathbf{A} = \begin{vmatrix} \mathbf{i} & \mathbf{j} & \mathbf{k} \\ \partial/\partial x & \partial/\partial y & \partial/\partial z \\ A_x & A_y & A_z \end{vmatrix}
$$

$$
= \mathbf{i}\left(\frac{\partial A_z}{\partial y} - \frac{\partial A_y}{\partial z}\right) + \mathbf{j}\left(\frac{\partial A_x}{\partial z} - \frac{\partial A_x}{\partial z}\right) + \mathbf{k}\left(\frac{\partial A_y}{\partial x} - \frac{\partial A_x}{\partial y}\right). \quad (12.83)
$$

The vector field $\nabla \times \mathbf{A}$ is called the *curl of* $\mathbf{A}$, also written curl $\mathbf{A}$. For an element of area in an arbitrary orientation, we find the limit

$$
\lim_{\Delta\sigma \to 0} \frac{\oint_{\Delta\sigma} \mathbf{A(r)} \cdot ds}{\Delta\sigma} = \nabla \times \mathbf{A}, \quad (12.84)
$$

where $\nabla \times \mathbf{A}$ is directed normal to the element of area. An arbitrary surface, such as the one drawn in Figure 12.15, can be formed from an infinite number of infinitesimal elements of area such as Figure 12.14. The line integral of a vector field $\mathbf{A(r)}$ counterclockwise around a path closed C can be related to surface integral of $\nabla \times \mathbf{A}$ over the enclosed area by *Stokes' theorem*:

$$
\oint_C \mathbf{A} \cdot d\mathbf{s} = \int_S (\nabla \times \mathbf{A}) \cdot d\boldsymbol{\sigma}. \quad (12.85)
$$

In two dimensions, Stokes' theorem reduces to Green's theorem (11.75):

$$\oint_C (A_x\, dx + A_y\, dy) = \int_S \left(\frac{\partial A_y}{\partial x} - \frac{\partial A_x}{\partial y} \right) dx\, dy. \tag{12.86}$$

For arbitrary scalar and vector fields $\phi(\mathbf{r})$ and $\mathbf{A}(\mathbf{r})$ the following identities involving divergence and curl can be readily derived:

$$\nabla \times \nabla\phi \equiv 0 \tag{12.87}$$

and

$$\nabla \cdot (\nabla \times \mathbf{A}) \equiv 0. \tag{12.88}$$

An intriguing combination is $\nabla \times (\nabla \times \mathbf{A})$ also known as "curl curl" (Curl Curl Beach is a suburban resort outside Sydney, Australia). This has the form of a triple vector product (12.38). We must be careful about the order of factors, however, since ∇ is a differential operator. If the vector $\mathbf{A}$ is always kept on the right of the ∇'s, we obtain the correct result:

$$\nabla \times (\nabla \times \mathbf{A}) = \nabla(\nabla \cdot \mathbf{A}) - \nabla^2 \mathbf{A}. \tag{12.89}$$

Note that $\nabla^2 \mathbf{A}$ is a vector quantity such that

$$\nabla^2 \mathbf{A} = \mathbf{i}\nabla^2 A_x + \mathbf{j}\nabla^2 A_y + \mathbf{k}\nabla^2 A_z. \tag{12.90}$$

12.11   Maxwell's Equations

These four fundamental equations of electromagnetic theory can be very elegantly expressed in terms of the vector operators, divergence and curl. The first Maxwell equation is a generalization of Coulomb's law for the electric field of a point charge:

$$\mathbf{E} = \frac{Q}{4\pi\epsilon_0 r^2}\hat{\mathbf{r}}, \tag{12.91}$$

where Q is the electric charge and ϵ_0, the permittivity of free space. The unit vector $\hat{\mathbf{r}}$ indicates that the field is directed radially outward from the charge (or radially inward for a negative charge). We write

$$4\pi r^2 \mathbf{E} \cdot \hat{\mathbf{r}} = \frac{Q}{\epsilon_0}. \tag{12.92}$$

Since $4\pi r^2 \hat{\mathbf{r}}$ represents an element of area with an outward-directed normal, this can be transformed into the integral relationship

$$\int_S \mathbf{E} \cdot d\boldsymbol{\sigma} = \frac{1}{\epsilon_0} \int_V \rho(\mathbf{r}) d\tau, \tag{12.93}$$

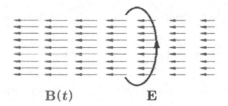

B(t)   E

Figure 12.16 Faraday's law of electromagnetic induction, leading to third Maxwell equation: $\nabla \times \mathbf{E} + \frac{\partial \mathbf{B}}{\partial t} = 0$.

where $\rho(\mathbf{r})$ is the charge density within a sphere of volume V with surface area S. But by the divergence theorem,

$$\int_S \mathbf{E} \cdot d\boldsymbol{\sigma} = \int_V \nabla \cdot \mathbf{E} \, d\tau. \tag{12.94}$$

The corresponding differential form is then the first of Maxwell's equations:

$$\nabla \cdot \mathbf{E} = \frac{\rho}{\epsilon_0} \quad (\text{Maxwell 1}). \tag{12.95}$$

Free magnetic poles, the magnetic analog of electric charges, have never been observed (although their existence has been postulated in some theories). This implies that the divergence of the magnetic induction $\mathbf{B}$ equals zero, which serves as the second Maxwell equation:

$$\nabla \cdot \mathbf{B} = 0 \quad (\text{Maxwell 2}). \tag{12.96}$$

According to Faraday's law of electromagnetic induction, a *time-varying* magnetic field induces an electromotive force in a conducting loop through which the magnetic field threads, as shown in Figure 12.16. This is somewhat suggestive of a linear relationship between the time derivative of the magnetic induction and the curl of the electric field. Indeed, the appropriate equation is

$$\nabla \times \mathbf{E} + \frac{\partial \mathbf{B}}{\partial t} = 0 \quad (\text{Maxwell 3}). \tag{12.97}$$

Oersted discovered that an electric current produces a magnetic field winding around it, as shown in Figure 12.17. The quantitative result, suggested by analogy with Eq. (12.97), is *Ampère's law*:

$$\nabla \times \mathbf{B} = \mu_0 \mathbf{J}, \tag{12.98}$$

where μ_0 is the permeability of free space. Maxwell showed that Ampère's law must be incomplete by the following argument. Taking the divergence of (12.98) we find

$$\nabla \cdot (\nabla \times \mathbf{B}) = \nabla \cdot \mathbf{J} = 0, \tag{12.99}$$

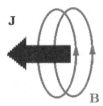

Figure 12.17 Magnetic field produced by electric current, described by Ampére's law: $\nabla \times \mathbf{B} = \mu_0 \mathbf{J}$.

since the divergence of a curl is identically zero [Eq. (12.88)]. This is consistent only for the special case of steady currents [Eq. (12.75)]. To make the right-hand side accord with the full equation of continuity, Eq. (12.74), we can write

$$\nabla \cdot (\nabla \times \mathbf{B}) = \frac{\partial \rho}{\partial t} + \nabla \cdot \mathbf{J}. \tag{12.100}$$

This is still equivalent to $0 = 0$ but suggests by (12.95) that

$$\nabla \cdot \frac{\partial \mathbf{E}}{\partial t} = \frac{1}{\epsilon_0} \frac{\partial \rho}{\partial t}. \tag{12.101}$$

With Maxwell's *displacement current hypothesis*, the generalization of Ampère's law is

$$\nabla \times \mathbf{B} = \mu_0 \mathbf{J} + \epsilon_0 \mu_0 \frac{\partial \mathbf{E}}{\partial t} \quad \text{(Maxwell 4)}, \tag{12.102}$$

the added term being known as the *displacement current*.

We have given the forms of Maxwell's equations in free space. In material media, two auxilliary fields are defined: the electric displacement $\mathbf{D} = \epsilon \mathbf{E}$ and the magnetic field $\mathbf{H} = \mathbf{B}/\mu$. In terms of these, we can write compact forms for Maxwell's equations in SI units:

$$\nabla \cdot \mathbf{D} = \rho, \quad \nabla \cdot \mathbf{B} = 0, \quad \nabla \times \mathbf{E} + \frac{\partial \mathbf{B}}{\partial t} = 0, \quad \nabla \times \mathbf{H} = \mathbf{J} + \frac{\partial \mathbf{D}}{\partial t}. \tag{12.103}$$

In the absence of charges and currents ($\rho = 0$ and $\mathbf{J} = \mathbf{0}$), Maxwell's equations reduce to

$$\nabla \cdot \mathbf{E} = 0, \quad \nabla \cdot \mathbf{B} = 0, \quad \nabla \times \mathbf{E} + \frac{\partial \mathbf{B}}{\partial t} = 0, \quad \nabla \times \mathbf{B} - \epsilon_0 \mu_0 \frac{\partial \mathbf{E}}{\partial t} = 0. \tag{12.104}$$

Taking the curl of the third equation, the time derivative of the fourth and eliminating the terms in $\mathbf{B}$, we find

$$\nabla \times (\nabla \times \mathbf{E}) + \epsilon_0 \mu_0 \frac{\partial^2 \mathbf{E}}{\partial t^2} = 0. \tag{12.105}$$

Using the curl curl equation (12.89) and noting that $\nabla \cdot \mathbf{E} = 0$ we obtain the *wave equation* for the electric field:

$$\nabla^2 \mathbf{E} - \frac{1}{c^2} \frac{\partial^2 \mathbf{E}}{\partial t^2} = 0. \tag{12.106}$$

An analogous derivation shows that the magnetic induction also satisfies the wave equation:

$$\nabla^2 \mathbf{B} - \frac{1}{c^2} \frac{\partial^2 \mathbf{B}}{\partial t^2} = 0. \tag{12.107}$$

Maxwell proposed that Eqs. (12.106) and (12.107) describe the propagation of *electromagnetic waves* at the speed of light

$$c = \frac{1}{\sqrt{\epsilon_0 \mu_0}}. \tag{12.108}$$

Note that this conclusion would not have been possible without the displacement current hypothesis.

By virtue of the vector identities (12.87) and (12.88), the two vector fields $\mathbf{E}$ and $\mathbf{B}$ can be represented by *electromagnetic potentials*—one vector field and one scalar field. The second Maxwell equation (12.96) suggests that the magnetic induction can be written

$$\mathbf{B} = \nabla \times \mathbf{A}, \tag{12.109}$$

where $\mathbf{A}(\mathbf{r}, t)$ is called the *vector potential*. Substituting this into the third Maxwell equation (12.97), we can write

$$\nabla \times \left(\mathbf{E} + \frac{\partial \mathbf{A}}{dt} \right) = 0. \tag{12.110}$$

This suggests that the quantity in parentheses can be represented as the divergence of a scalar function, conventionally written in the form

$$\mathbf{E} + \frac{\partial \mathbf{A}}{dt} = -\nabla \Phi, \tag{12.111}$$

where $\Phi(\mathbf{r}, t)$ is called the *scalar potential*. In the time-independent case, the latter reduces to the Coulomb or electrostatic potential, with $\mathbf{E} = -\nabla \Phi$. The electric field and magnetic induction are uniquely determined by the scalar and vector potentials:

$$\mathbf{E} = -\nabla \Phi - \frac{\partial \mathbf{A}}{dt} \quad \text{and} \quad \mathbf{B} = \nabla \times \mathbf{A}. \tag{12.112}$$

The converse is not true, however. Consider the alternative choice of electromagnetic potentials

$$\mathbf{A}' = \mathbf{A} + \nabla \chi, \quad \Phi' = \Phi - \frac{\partial \chi}{\partial t}, \tag{12.113}$$

where $\chi(\mathbf{r}, t)$ is an arbitrary function. The modified potentials $\mathbf{A}'$ and Φ' can be verified to give the *same* $\mathbf{E}$ and $\mathbf{B}$ as the original potentials. This property of electromagnetic fields is called *gauge invariance*. Extension of this principle to quantum theory leads to the concept of *gauge fields*, which provides the framework of the standard model for elementary particles and their interactions.

12.12 Covariant Electrodynamics

Electromagnetic theory can be very compactly expressed in Minkowski four-vector notation. Historically, it was this symmetry of Maxwell's equations which led to the Special Theory of Relativity. Now that we know about partial derivatives, the *gradient* four-vector can be defined. This is a covariant operator

$$\partial_\mu = \left[\partial/\partial x^0 \; \partial/\partial x^1 \; \partial/\partial x^2 \; \partial/\partial x^3 \right] = \left[c^{-1}\partial/\partial t \; \partial/\partial x \; \partial/\partial y \; \partial/\partial z \right]. \tag{12.114}$$

The corresponding contravariant operator is

$$\partial^\mu = \begin{bmatrix} \partial/\partial x^0 \\ -\partial/\partial x^1 \\ -\partial/\partial x^2 \\ -\partial/\partial x^3 \end{bmatrix} = \begin{bmatrix} c^{-1}\partial/\partial t \\ -\partial/\partial x \\ -\partial/\partial y \\ -\partial/\partial z \end{bmatrix}. \tag{12.115}$$

The scalar product gives

$$\partial_\mu \partial^\mu = \frac{1}{c^2}\frac{\partial^2}{\partial t^2} - \frac{\partial^2}{\partial x^2} - \frac{\partial^2}{\partial y^2} - \frac{\partial^2}{\partial z^2} = \frac{1}{c^2}\frac{\partial^2}{\partial t^2} - \nabla^2. \tag{12.116}$$

The *D'Alembertian operator* $\square$ is defined by

$$\square \equiv \nabla^2 - \frac{1}{c^2}\frac{\partial^2}{\partial t^2}, \tag{12.117}$$

a notation is suggested by analogy with the symbol Δ for the Laplacian operator ∇^2. The wave equations (12.106) and (12.107) can thus be compactly written

$$\square \mathbf{E} = 0 \quad \text{and} \quad \square \mathbf{B} = 0. \tag{12.118}$$

For the Minkowski signature $\{+ - --\}$, $\partial_\mu \partial^\mu = -\square$. The alternative choice of signature $\{- + ++\}$ implies $\partial_\mu \partial^\mu = \square$. To confuse matters further, $\square^2$ is sometimes written in place of $\square$.

The equation of continuity for electric charge (12.74) has the structure of a four-dimensional divergence

$$\partial_\mu J^\mu = \partial^\mu J_\mu = 0, \tag{12.119}$$

in terms of the *charge-current four-vector*

$$J^\mu = \begin{bmatrix} c\rho \\ J_x \\ J_y \\ J_z \end{bmatrix}. \tag{12.120}$$

In *Lorenz gauge*, the function $\chi(\mathbf{r}, t)$ in Eqs. (12.113) is chosen such that $\Box \chi = 0$. The scalar and vector potentials are then related by the *Lorenz condition*

$$\nabla \cdot \mathbf{A} + \frac{1}{c^2} \frac{\partial \Phi}{\partial t} = 0. \tag{12.121}$$

This can also be expressed as a four-dimensional divergence

$$\partial_\mu A^\mu = \partial^\mu A_\mu = 0, \tag{12.122}$$

in terms of a four-vector potential

$$A^\mu = \begin{bmatrix} \Phi \\ A_x \\ A_y \\ A_z \end{bmatrix}. \tag{12.123}$$

Maxwell's equations in Lorenz gauge can be expressed by a single Minkowski-space equation

$$\Box A^\mu = \frac{4\pi}{c} J^\mu, \tag{12.124}$$

where the $\mu = 0$ component gives

$$\Box \Phi = 4\pi \rho. \tag{12.125}$$

(Incidentally, the Lorenz gauge was proposed by the Danish physicist Ludvig Lorenz. It is often erroneously designated "Lorentz gauge," after the more famous Dutch physicist Hendrik Lorentz. In fact, the condition does fulfill the property known as *Lorentz invariance*.)

The electric and magnetic fields obtained from the potentials using (12.112) can be represented by a single relation for the *electromagnetic field tensor*

$$F^{\mu\nu} = \partial^\mu A^\nu - \partial^\nu A^\mu = \begin{bmatrix} 0 & E_x & E_y & E_z \\ -E_x & 0 & -B_z & B_y \\ -E_y & B_z & 0 & -B_x \\ -E_z & -B_y & B_x & 0 \end{bmatrix}. \tag{12.126}$$

Maxwell's equations, in terms of the electromagnetic field tensor, can be written:

$$\partial_\mu F^{\mu\nu} = \frac{4\pi}{c} J^\nu \quad \text{and} \quad \partial_\lambda F_{\mu\nu} + \partial_\mu F_{\nu\lambda} + \partial_\nu F_{\lambda\mu} = 0. \tag{12.127}$$

Problem 12.12.1. An electromagnetic field *dual tensor* can be defined by

$$\mathcal{F}^{\lambda\mu} = \frac{1}{2}\epsilon^{\lambda,\mu,\alpha,\beta} F_{\alpha,\beta},$$

where $\epsilon^{\lambda,\mu,\alpha,\beta}$ is the four-dimensional Levi-Civita tensor. Verify that

$$\mathcal{F}^{\mu\nu} = \begin{bmatrix} 0 & -B_x & -B_y & -B_z \\ B_x & 0 & E_x & -E_y \\ B_y & -E_z & 0 & E_x \\ B_z & E_y & -E_x & 0 \end{bmatrix}.$$

This can be obtained from the $F^{\mu\nu}$ tensor by the formal substitutions $\mathbf{E} \to \mathbf{B}$ and $\mathbf{B} \to -\mathbf{E}$.

Problem 12.12.2. Show that the second of Maxwell's equations in (12.127) can be written more compactly as

$$\partial_\mu \mathcal{F}^{\mu,\nu} = 0.$$

12.13 Curvilinear Coordinates

Vector equations can provide an elegant abstract formulation of physical laws, but, in order to solve problems, it is usually necessary to express these equations in a particular coordinate system. Thus far we have considered Cartesian coordinates almost exclusively. Another choice, for example, cylindrical or spherical coordinates, might prove more appropriate, depending on the symmetry of the problem.

We will consider the general class of *orthogonal curvilinear coordinates*, designated q_1, q_2, q_3, whose coordinate surfaces always mutually intersect at right angles. The Cartesian coordinates of a point in three-dimensional space can be expressed in terms of a set of curvilinear coordinates by relations of the form

$$x = x(q_1, q_2, q_3), \quad y = y(q_1, q_2, q_3), \quad z = z(q_1, q_2, q_3). \tag{12.128}$$

The differentials of the Cartesian coordinates are

$$dx = \frac{\partial x}{\partial q_1}dq_1 + \frac{\partial x}{\partial q_2}dq_2 + \frac{\partial x}{\partial q_3}dq_3, \tag{12.129}$$

with analogous expressions for dy and dz. Thus a differential element of displacement $d\mathbf{r} = \mathbf{i}\,dx + \mathbf{j}\,dy + \mathbf{k}\,dz$ can be written

$$d\mathbf{r} = \frac{\partial \mathbf{r}}{\partial q_1}dq_1 + \frac{\partial \mathbf{r}}{\partial q_2}dq_2 + \frac{\partial \mathbf{r}}{\partial q_3}dq_3 = \hat{\mathbf{e}}_1 Q_1\,dq_1 + \hat{\mathbf{e}}_2 Q_2\,dq_2 + \hat{\mathbf{e}}_3 Q_3\,dq_3, \tag{12.130}$$

where $\hat{\mathbf{e}}_1, \hat{\mathbf{e}}_2, \hat{\mathbf{e}}_3$ are unit vectors with respect to the curvilinear coordinates q_1, q_2, q_3 and Q_1, Q_2, Q_3 are *scale factors*. As shown in Figure 12.18, the elements of length in the three coordinate directions are equal to $Q_1 \, dq_1$, $Q_2 \, dq_2$, and $Q_3 \, dq_3$. The element of volume is evidently given by

$$d\tau = Q_1 Q_2 Q_3 \, dq_1 \, dq_2 \, dq_3, \qquad (12.131)$$

where the Q_i are, in general, functions of q_1, q_2, q_3.

We can evidently identify

$$\hat{\mathbf{e}}_i \, Q_i = \frac{\partial \mathbf{r}}{\partial q_i}, \quad i = 1, 2, 3, \qquad (12.132)$$

so that the element of volume can be equated to a triple scalar product (cf Section 12.4):

$$d\tau = (\hat{\mathbf{e}}_1 Q_1 \, dq_1) \cdot (\hat{\mathbf{e}}_2 Q_2 \, dq_2) \times (\hat{\mathbf{e}}_3 Q_3 \, dq_3)$$
$$= \left(\frac{d\mathbf{r}}{dq_1}\right) \cdot \left(\frac{d\mathbf{r}}{dq_2}\right) \times \left(\frac{d\mathbf{r}}{dq_3}\right) dq_1 \, dq_2 \, dq_3. \qquad (12.133)$$

This can therefore be connected to the Jacobian determinant:

$$d\tau = \begin{vmatrix} \partial x/\partial q_1 & \partial x/\partial q_2 & \partial x/\partial q_3 \\ \partial y/\partial q_1 & \partial y/\partial q_2 & \partial y/\partial q_3 \\ \partial z/\partial q_1 & \partial z/\partial q_2 & \partial z/\partial q_3 \end{vmatrix} dq_1 \, dq_2 \, dq_3 = \frac{\partial(x, y, z)}{\partial(q_1, q_2, q_3)} dq_1 \, dq_2 \, dq_3. \qquad (12.134)$$

The components of the gradient vector represent directional derivatives of a function. For example, the change in the function $\phi(q_1, q_2, q_3)$ along the q_i-direction is given by the ratio of $d\phi$ to the element of length $Q_i \, dq_i$. Thus, the gradient in curvilinear coordinates can be written

$$\nabla \phi = \frac{\hat{\mathbf{e}}_1}{Q_1} \frac{\partial \phi}{\partial q_1} + \frac{\hat{\mathbf{e}}_2}{Q_2} \frac{\partial \phi}{\partial q_2} + \frac{\hat{\mathbf{e}}_3}{Q_3} \frac{\partial \phi}{\partial q_3}. \qquad (12.135)$$

The divergence $\nabla \cdot \mathbf{A}$ represents the limiting value of the net outward flux of the vector quantity $\mathbf{A}$ per unit volume. Referring to Figure 12.19, the net flux of the component A_1 in the q_1-direction is given by the difference between the *outward* contributions $Q_2 Q_3 A_1 \, dq_2 \, dq_3$ on the two shaded faces. As the volume element approaches a point, this reduces to

$$\frac{\partial(Q_2 Q_3 A_1)}{\partial q_1} dq_1 \, dq_2 \, dq_3. \qquad (12.136)$$

Adding the analogous contributions from the q_2- and q_3-directions and diving by the volume $d\tau$, we obtain the general result for the divergence in curvilinear coordinates:

$$\nabla \cdot \mathbf{A} = \frac{1}{Q_1 Q_2 Q_3} \left[\frac{\partial}{\partial q_1} Q_2 Q_3 A_1 + \frac{\partial}{\partial q_2} Q_3 Q_1 A_2 + \frac{\partial}{\partial q_3} Q_1 Q_2 A_3 \right]. \qquad (12.137)$$

The curl in curvilinear coordinates is given by

$$\nabla \times \mathbf{A} = \frac{1}{Q_1 Q_2 Q_3} \begin{vmatrix} Q_1 \hat{\mathbf{e}}_1 & Q_2 \hat{\mathbf{e}}_2 & Q_3 \hat{\mathbf{e}}_3 \\ \partial/\partial q_1 & \partial/\partial q_2 & \partial/\partial q_3 \\ Q_1 A_1 & Q_2 A_2 & Q_3 A_3 \end{vmatrix}. \qquad (12.138)$$

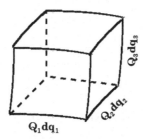

Figure 12.18 Volume element in curvilinear coordinates q_1, q_2, q_3: $d\tau = Q_1 Q_2 Q_3 \, dq_1 \, dq_2 \, dq_3$.

The Laplacian is the divergence of the gradient, so that substitution of (12.135) into (12.137) gives

$$\nabla^2 = \frac{1}{Q_1 Q_2 Q_3} \left[\frac{\partial}{\partial q_1} \frac{Q_2 Q_3}{Q_1} \frac{\partial}{\partial q_1} + \frac{\partial}{\partial q_2} \frac{Q_3 Q_1}{Q_2} \frac{\partial}{\partial q_2} + \frac{\partial}{\partial q_3} \frac{Q_1 Q_2}{Q_3} \frac{\partial}{\partial q_3} \right].$$
(12.139)

The three most common coordinate systems in physical applications are Cartesian, with $Q_x = Q_y = Q_z = 1$, cylindrical, with $Q_r = 1$, $Q_\theta = r$, $Q_z = 1$, and spherical polar, with $Q_r = 1$, $Q_\theta = r$, $Q_\phi = r \sin \theta$. We frequently encounter the spherical polar volume element

$$d\tau = r^2 \sin \theta \, dr \, d\theta \, d\phi$$
(12.140)

and the Laplacian operator

$$\nabla^2 = \frac{1}{r^2} \frac{\partial}{\partial r} r^2 \frac{\partial}{\partial r} + \frac{1}{r^2 \sin \theta} \frac{\partial}{\partial \theta} \sin \theta \frac{\partial}{\partial \theta} + \frac{1}{r^2 \sin^2 \theta} \frac{\partial^2}{\partial \phi^2}.$$
(12.141)

Problem 12.13.1. Determine the forms of the scalar and vector products, $\mathbf{A} \cdot \mathbf{B}$ and $\mathbf{A} \times \mathbf{B}$, in terms of the components of $\mathbf{A}$ and $\mathbf{B}$ in spherical polar coordinates.

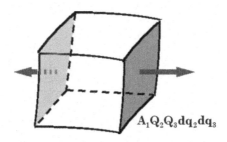

Figure 12.19 Evaluation of divergence in curvilinear coordinates.

12.14   Vector Identities

For ready reference, we list a selection of useful vector identities.

$$\mathbf{A} \cdot (\mathbf{B} \times \mathbf{C}) = \mathbf{B} \cdot (\mathbf{C} \times \mathbf{A}) = \mathbf{C} \cdot (\mathbf{A} \times \mathbf{B}),$$

$$\mathbf{A} \times (\mathbf{B} \times \mathbf{C}) = \mathbf{B}(\mathbf{A} \cdot \mathbf{C}) - \mathbf{C}(\mathbf{A} \cdot \mathbf{B}),$$

$$(\mathbf{A} \times \mathbf{B}) \cdot (\mathbf{C} \times \mathbf{D}) = (\mathbf{A} \cdot \mathbf{C})(\mathbf{B} \cdot \mathbf{D}) - (\mathbf{A} \cdot \mathbf{D})(\mathbf{B} \cdot \mathbf{C}),$$

$$\nabla(fg) = f\nabla g + g\nabla f,$$

$$\nabla \cdot (f\mathbf{A}) = \mathbf{A} \cdot (\nabla f) + f(\nabla \cdot \mathbf{A}),$$

$$\nabla \times (f\mathbf{A}) = (\nabla f) \times \mathbf{A} + f(\nabla \times \mathbf{A}),$$

$$\nabla(\mathbf{A} \cdot \mathbf{B}) = \mathbf{B} \times (\nabla \times \mathbf{A}) + \mathbf{A} \times (\nabla \times \mathbf{B}) + (\mathbf{B} \cdot \nabla)\mathbf{A} + (\mathbf{A} \cdot \nabla)\mathbf{B},$$

$$\nabla \cdot (\mathbf{A} \times \mathbf{B}) = \mathbf{B} \cdot (\nabla \times \mathbf{A}) - \mathbf{A} \cdot (\nabla \times \mathbf{B}),$$

$$\nabla \times (\mathbf{A} \times \mathbf{B}) = \mathbf{A}(\nabla \cdot \mathbf{B}) + (\mathbf{B} \cdot \nabla)\mathbf{A} - \mathbf{B}(\nabla \cdot \mathbf{A}) - (\mathbf{A} \cdot \nabla)\mathbf{B},$$

$$\nabla \times (\nabla \times \mathbf{A}) = \nabla(\nabla \cdot \mathbf{A}) - \nabla^2\mathbf{A}.$$

Problem 12.14.1.   Derive some of the above vector identities. It is usually sufficient to demonstrate a single Cartesian component of the formula.

13 Partial Differential Equations and Special Functions

13.1 Partial Differential Equations

Many complex physical, geometrical, and stochastic (probabilistic) phenomena are described by *partial differential equations* (PDEs), involving two or more independent variables. They are obviously much more difficult to solve than ordinary differential equations (ODEs). Some applications, including weather prediction, econometric models, fluid dynamics, and nuclear engineering, might involve simultaneous PDEs with large numbers of variables. Such problems are best tackled by powerful supercomputers. We will be content to consider a few representative second-order PDEs for which analytic solutions are possible.

Scalar and vector fields which depend on more than one independent variable, which we write in the notation $\Psi(x, y)$, $\Psi(x, t)$, $\Psi(\mathbf{r})$, $\Psi(\mathbf{r}, t)$, etc., are very often obtained as solutions to PDEs. Some classic equations of mathematical physics which we will consider are the wave equation, the heat equation, Laplace's equation, Poisson's equation, and the Schrödinger equation for some exactly solvable quantum-mechanical problems.

Maxwell's equations lead to the wave equation for components of the electric and magnetic fields, with the general form

$$\nabla^2 \Psi - \frac{1}{c^2} \frac{\partial^2 \Psi}{\partial t^2} = 0. \tag{13.1}$$

Analogous equations apply to sound and other wave phenomena. For waves in one spatial dimension, the wave equation reduces to

$$\frac{\partial^2 \Psi}{\partial x^2} - \frac{1}{c^2} \frac{\partial^2 \Psi}{\partial t^2} = 0. \tag{13.2}$$

The *heat equation* or *diffusion equation* is similar to the wave equation but with a first derivative in the time variable:

$$\frac{\partial \Psi}{\partial t} = \kappa \nabla^2 \Psi, \tag{13.3}$$

where κ is a constant determined by the thermal conductivity or diffusion coefficient.

Guide to Essential Math 2e. http://dx.doi.org/10.1016/B978-0-12-407163-6.00013-8

Introducing the electrostatic potential $\Phi(\mathbf{r}, t)$ into the first of Maxwell's equation, $\nabla \cdot \mathbf{E} = \rho/\epsilon_0$, we obtain *Poisson's equation*

$$\nabla^2 \Phi = -\rho(\mathbf{r})/\epsilon_0. \tag{13.4}$$

In a region free of electric charge, this reduces to Laplace's equation

$$\nabla^2 \Phi = 0. \tag{13.5}$$

In the calculus of complex variables, we encounter the two-dimensional Laplace equation for $u(x, y)$

$$\frac{\partial^2 u}{\partial x^2} + \frac{\partial^2 u}{\partial y^2} = 0. \tag{13.6}$$

For the case of monochromatic time dependence, the solution of the wave equation has the form

$$\Psi(\mathbf{r}, t) = \psi(\mathbf{r})T(t), \tag{13.7}$$

where

$$T(t) = \begin{cases} \sin \omega t, \\ \cos \omega t, \end{cases} \tag{13.8}$$

or some linear combination of sine and cosine or complex exponentials. Substituting (13.7) into the wave equation (13.1) we find

$$T(t)\nabla^2 \psi(\mathbf{r}) - \psi(\mathbf{r})\frac{T''(t)}{c^2} = 0, \tag{13.9}$$

where we have noted that ∇^2 acts only on the factor $\psi(\mathbf{r})$ while $\partial^2/\partial t^2$ acts only on $T(t)$. Since either of the functions (13.8) is a solution of the ordinary differential equation

$$T''(t) + \omega^2 T(t) = 0. \tag{13.10}$$

Equation (13.9) reduces to *Helmholtz's equation*

$$\nabla^2 \psi(\mathbf{r}) + k^2 \psi(\mathbf{r}) = 0 \tag{13.11}$$

after canceling out $T(t)$ and defining $k \equiv \omega/c$. The heat equation also reduces to Helmholtz's equation for separable time dependence in the form

$$T(t) = e^{-\kappa t}. \tag{13.12}$$

Recall that ordinary differential equations generally yield solutions containing one or more arbitrary constants, which can be determined from boundary conditions. In contrast, solutions of partial differential equations often contain arbitrary *functions*. An extreme case is the second-order equation

$$\frac{\partial^2 F(x, y)}{\partial x \partial y} = F_{x,y}(x, y) = 0. \tag{13.13}$$

This has the solutions

$$F(x, y) = f_1(x) + f_2(y),$$ (13.14)

where f_1 and f_2 are arbitrary differentiable functions. More detailed boundary conditions must be specified in order to find more specific solutions.

13.2 Separation of Variables

The simplest way to solve PDEs is to reduce a PDE with n independent variables to n independent ODEs, each depending on just one variable. This isn't always possible but we will limit our consideration to such cases. In the preceding section, we were able to reduce the wave equation and the heat equation to the Helmholtz equation when the time dependence of $\Psi(\mathbf{r}, t)$ was separable. Consider the Helmholtz equation in Cartesian coordinates

$$(\nabla^2 + k^2)\Psi(x, y, z) = \frac{\partial^2 \Psi}{\partial x^2} + \frac{\partial^2 \Psi}{\partial y^2} + \frac{\partial^2 \Psi}{\partial z^2} + k^2\Psi(x, y, z) = 0.$$ (13.15)

If the boundary conditions allow, the solution Ψ might be reducible to a *separable* function of x, y, z:

$$\Psi(x, y, z) = X(x)Y(y)Z(z).$$ (13.16)

Substituting this into (13.15), we obtain

$$X''(x)Y(y)Z(z) + X(x)Y''(y)Z(z) + X(x)Y(y)Z''(z) + k^2X(x)Y(y)Z(z) = 0.$$ (13.17)

Division by $X(x)Y(y)Z(z)$ gives the simplified equation:

$$\frac{X''(x)}{X(x)} + \frac{Y''(y)}{Y(y)} + \frac{Z''(z)}{Z(z)} + k^2 = 0.$$ (13.18)

Now, if we solve for the term $X''(x)/X(x)$, we find this function of x alone must be equal to a function of y and z, for arbitrary values of x, y, z. The only way this is possible is for the function to equal a constant, say

$$\frac{X''(x)}{X(x)} = -\alpha^2.$$ (13.19)

A negative constant is consistent with a periodic function $X(x)$, otherwise it would have exponential dependence. Analogously,

$$\frac{Y''(y)}{Y(y)} = -\beta^2 \quad \text{and} \quad \frac{Z''(z)}{Z(z)} = -\gamma^2,$$ (13.20)

so that

$$\alpha^2 + \beta^2 + \gamma^2 = k^2. \tag{13.21}$$

We have thereby reduced Eq. (13.15) to three ordinary differential equations:

$$X''(x) + \alpha^2 X(x) = 0, \quad Y''(y) + \beta^2 Y(y) = 0, \quad Z''(z) + \gamma^2 Z(z) = 0. \tag{13.22}$$

We solved an equation of this form in Section 8.5, for two alternative sets of boundary conditions. The preceding solution of the Helmholtz equation in Cartesian coordinates is applicable to the Schrödinger equation for the quantum-mechanical particle-in-a-box problem.

13.3 Special Functions

The designation *elementary functions* is usually applied to the exponential, logarithmic, and trigonometric functions and combinations formed by algebraic operations. Certain other *special functions* occur so frequently in applied mathematics that they acquire standardized symbols and names—often in honor of a famous mathematician. We will discuss a few examples of interest in physics, chemistry, and engineering, in particular, the special functions named after Bessel, Legendre, Hankel, Laguerre, and Hermite. We already encountered in Chapter 6 the gamma function, error function, and exponential integral. Special functions are most often solutions of second-order ODEs with variable coefficients, obtained after separation of variables in PDEs. So that you will have seen the names, here is a list of some other special functions you are likely to encounter (although not in this book): Airy functions, Beta functions, Chebyshev polynomials, Elliptic functions, Gegenbauer polynomials, Jacobi polynomials, Mathieu functions, Meijer G-functions, Parabolic cylinder functions, Theta functions, Whittaker functions.

For a comprehensive reference on special functions, see M. Abramowitz and I. A. Stegun, Eds., *Handbook of Mathematical Functions with Formulas, Graphs, and Mathematical Tables*, (National Bureau of Standards Applied Mathematics Series, Vol. 55, Washington, DC, 1964). This is now available online at http://dlmf.nist.gov and features graphics and hypertext links.

Following is a sampling of some "amazing" formulas involving gamma and zeta functions, two of the special functions we have already introduced. A reflection formula for gamma functions:

$$\Gamma(1 + iz)\Gamma(1 - iz) = \frac{\pi z}{\sinh \pi z}. \tag{13.23}$$

An infinite product representation of the gamma function:

$$\frac{1}{\Gamma(z)} = z e^{\gamma z} \prod_{n=1}^{\infty} \left(1 + \frac{z}{n}\right) e^{-z/n}, \tag{13.24}$$

where $\gamma = 0.5772\ldots$, the Euler-Mascheroni constant. An integral for the Riemann zeta function:

$$\zeta(s) = \frac{1}{\Gamma(s)} \int_0^\infty \frac{x^{s-1}}{e^x - 1} dx. \tag{13.25}$$

Euler's relation connecting the zeta function with prime numbers, which we proved in Section 1.12:

$$\zeta(s) = \prod_p \left(1 - p^{-s}\right)^{-1}. \tag{13.26}$$

For a long time, I felt intimidated by these amazing formulas involving special functions. Some resembled near miracles like "pulling rabbits out of a hat." I could never, in a thousand years, have come up with anything as brilliant as one of these myself. I would not have been surprised to see in some derivation "and then a miracle occurs" between the next-to-last and last equations (see Sidney Harris cartoon):

"I THINK YOU SHOULD BE MORE EXPLICIT HERE IN STEP TWO."

Over the years, I learned to be less intimidated, assuring myself these amazing results were the product of hundreds of brilliant mathematicians using inspired guesswork, reverse engineering, and all kinds of other nefarious practices for over 300 years. Ideally, one should be fully comfortable in using these results, resisting any thoughts that we are not entitled to exploit such brilliance. Something like the habitual comfort most of us have developed in using our cars, computers, and washing machines. Without a doubt, we all "stand on the shoulders of giants."

13.4   Leibniz's Formula

We will later need a formula for the nth derivative of the product of two functions. This can be derived stepwise as follows:

$$\frac{d}{dx}\big[f(x)g(x)\big] = f'(x)g(x) + f(x)g'(x),$$

$$\frac{d^2}{dx^2}\big[f(x)g(x)\big] = f''(x)g(x) + 2f'(x)g'(x) + f(x)g''(x),$$

$$\frac{d^3}{dx^3}\big[f(x)g(x)\big] = f'''(x)g(x) + 3f''(x)g'(x) + 3f'(x)g''(x) + f(x)g'''(x), \ldots$$

$$(13.27)$$

Clearly, we are generating a series containing the binomial coefficients

$$\binom{n}{m} = \frac{n!}{m!(n-m)!}, \quad m = 0, 1, \ldots, n \tag{13.28}$$

and the general result is Leibniz's formula

$$\frac{d^n}{dx^n}\big[f(x)g(x)\big] = \sum_{m=0}^{n} \binom{n}{m} f^{(n-m)}(x)g^{(m)}(x)$$

$$= f^{(n)}(x)g(x) + nf^{(n-1)}(x)g'(x)$$

$$+ \frac{n(n-1)}{2} f^{(n-2)}(x)g''(x) + \cdots . \tag{13.29}$$

13.5   Vibration of a Circular Membrane

As our first example of a special function, we consider a two-dimensional problem: the vibration of a circular membrane such as a drumhead. The amplitude of vibration is determined by solution of the Helmholtz equation in two dimensions, most appropriately chosen as the polar coordinates r, θ. Using the scale factors $Q_r = 1$ and $Q_\theta = r$ for the two-dimensional Laplacian, the Helmholtz equation can be written

$$(\nabla^2 + k^2)\Psi(r, \theta) = \frac{1}{r}\frac{\partial}{\partial r}\left(r\frac{\partial \Psi}{\partial r}\right) + \frac{1}{r^2}\frac{\partial^2 \Psi}{\partial \theta^2} + k^2\Psi(r, \theta) = 0. \tag{13.30}$$

This is subject to the boundary condition that the rim of the membrane at $r = r_0$ is fixed, so that $\Psi(r_0, \theta) = 0$. Also it is necessary that the amplitude be finite everywhere on the membrane, in particular, at $r = 0$. Assuming a separable solution $\Psi(r, \theta) = R(r)\Theta(\theta)$, Helmholtz's equation reduces to

$$R''(r)\Theta(\theta) + r^{-1}R'(r)\Theta(\theta) + r^{-2}R(r)\Theta''(\theta) + k^2 R(r)\Theta(\theta) = 0 \tag{13.31}$$

or

$$R''(r) + r^{-1} R'(r) + r^{-2} R(r) \left[\frac{\Theta''(\theta)}{\Theta(\theta)} \right] + k^2 R(r) = 0. \tag{13.32}$$

We could complete a separation of variables by dividing by $r^{-2} R(r)$. This would imply that the function of θ in square brackets equals a constant, which we can write

$$\frac{\Theta''(\theta)}{\Theta(\theta)} = -m^2 \Rightarrow \Theta''(\theta) + m^2 \Theta(\theta) = 0. \tag{13.33}$$

This is a familiar equation, with linearly independent solutions

$$\Theta(\theta) = \begin{cases} \sin m\theta, \\ \cos m\theta. \end{cases} \tag{13.34}$$

Since θ is an angular variable, the periodicity $\Theta(\theta \pm 2n\pi) = \Theta(\theta)$ is required. This restricts the parameter m to integer values. Thus we obtain an ordinary differential equation for $R(r)$:

$$R''(r) + r^{-1} R'(r) - m^2 r^{-2} R(r) + k^2 R(r) = 0. \tag{13.35}$$

Changing the variable to $x \equiv kr$, with $J(x) \equiv R(r)$, we can write

$$x^2 J''(x) + x J'(x) + (x^2 - m^2) J(x) = 0. \tag{13.36}$$

This is recognized as Bessel's equation (8.135). Only the Bessel functions $J_m(kr)$ are finite at $r = 0$, the Neumann functions $Y_m(kr)$ being singular there. Thus the solutions to Helmholtz's equation are

$$\Psi(r, \theta) = J_m(kr) \begin{cases} \sin m\theta, & m = 1, 2, 3, \dots, \\ \cos m\theta, & m = 0, 1, 2, \dots \end{cases} \tag{13.37}$$

The boundary condition at $r = r_0$ requires that

$$J_m(kr_0) = 0. \tag{13.38}$$

The zeros of Bessel functions are extensively tabulated. Let x_{mn} represent the nth root of $J_m(x) = 0$. The first three zeros are given by:

n	x_{0n}	x_{1n}	x_{2n}
1	2.4048	3.8317	5.1356
2	5.5201	7.0156	8.4172
3	8.6537	10.1735	11.6198

The eigenvalues of k are given by

$$k_{mn} = x_{mn}/r_0. \tag{13.39}$$

Some modes of vibration of a circular membrane, as labeled by the quantum numbers m and n, are sketched in Figure 13.1. To simplify the figure, only the sign of the wavefunction is indicated: gray for positive, white for negative. Note that modes for $m > 0$ are twofold degenerate, corresponding to the factors $\cos m\theta$ and $\sin m\theta$.

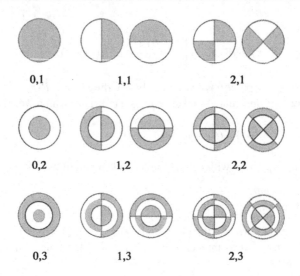

0,1   1,1   2,1

0,2   1,2   2,2

0,3   1,3   2,3

Figure 13.1 Modes of vibration (m, n) of a circular membrane. Wavefunctions are positive in gray areas, negative in white.

13.6 Bessel Functions

In Section 8.7, a series solution of Bessel's differential equation was derived, leading to the definition of *Bessel functions of the first kind*:

$$J_n(x) = \left(\frac{x}{2}\right)^n \left[1 - \frac{1}{(n+1)!}\left(\frac{x}{2}\right)^2 + \frac{1}{2!(n+2)!}\left(\frac{x}{2}\right)^4 - \cdots\right]$$

$$= \sum_{k=0}^{\infty} \frac{(-1)^k}{k!(n+k)!}\left(\frac{x}{2}\right)^{n+2k}. \tag{13.40}$$

This applies as well for noninteger n with the replacements $(n+k)! \rightarrow \Gamma(n+k+1)$. The Bessel functions are normalized in the sense that

$$\int_0^\infty J_n(x)dx = 1. \tag{13.41}$$

Bessel functions of integer order can be determined from expansion of a *generating function*:

$$\exp\left[\frac{x}{2}\left(t - \frac{1}{t}\right)\right] = \sum_{n=-\infty}^{\infty} J_n(x)t^n. \tag{13.42}$$

To see how this works, expand the product of the two exponentials:

$$e^{xt/2}e^{-x/2t} = \left[1 + \frac{x}{2}t + \frac{1}{2!}\left(\frac{x}{2}\right)^2 t^2 + \cdots\right]\left[1 - \frac{x}{2}t^{-1} + \frac{1}{2!}\left(\frac{x}{2}\right)^2 t^{-2} - \cdots\right].$$

(13.43)

The terms which contribute to t^0 are

$$1 - \left(\frac{x}{2}\right)^2 + \frac{1}{(2!)^2}\left(\frac{x}{2}\right)^4 - \cdots,$$

(13.44)

which gives the expansion for $J_0(x)$. The expansion for $J_1(x)$ is found from the terms proportional to t^1 and so forth.

One of the many integral representations of Bessel functions is

$$J_n(x) = \frac{1}{\pi}\int_0^\pi \cos(z\sin\theta - n\theta)d\theta.$$

(13.45)

An identity involving the derivative is

$$\frac{d}{dx}\left[x^n J_n(x)\right] = x^n J_{n-1}(x).$$

(13.46)

In particular, $J_0'(x) = -J_1(x)$. An addition theorem for Bessel functions states that

$$J_n(x+y) = \sum_{m=0}^\infty J_n(x)J_{m-n}(y).$$

(13.47)

Asymptotic forms for the Bessel function for small and large values of the argument are given by

$$J_n(x) \approx \frac{1}{n!}\left(\frac{x}{2}\right)^n \quad \text{for} \quad x \to 0$$

(13.48)

and

$$J_n(x) \approx \sqrt{\frac{2}{\pi x}}\cos\left[x - (2n+1)\frac{\pi}{4}\right] \quad \text{for} \quad x \gg n.$$

(13.49)

Bessel functions of the second kind show the following limiting behavior:

$$Y_0(x) \approx \frac{2}{\pi}\ln x, \quad Y_n(x) \approx -\frac{2^n(n-1)!}{\pi}x^{-n} \quad (n>0) \quad \text{for} \quad x \to 0$$

(13.50)

and

$$Y_n(x) \approx \sqrt{\frac{2}{\pi x}}\sin\left[x - (2n+1)\frac{\pi}{4}\right] \quad \text{for} \quad x \gg n.$$

(13.51)

The two kinds of Bessel functions thus have the asymptotic dependence of slowly damped cosines and sines. In analogy with Euler's formula $e^{\pm ix} = \cos x \pm i \sin x$, we define *Hankel functions* of the first and second kind:

$$H_n^{(1)}(x) = J_n(x) + iY_n(x) \quad \text{and} \quad H_n^{(2)}(x) = J_n(x) - iY_n(x). \tag{13.52}$$

The asymptotic forms of the Hankel functions are given by

$$H_n^{(1,2)}(x) \approx \sqrt{\frac{2}{\pi x}} e^{\pm i[x - (2n+1)\pi/4]} \quad \text{for} \quad x \gg n \tag{13.53}$$

having the character of waves propagating to the right and left, respectively.

Trigonometric functions with imaginary arguments suggested the introduction of hyperbolic functions: $\sinh x = -i \sin ix$ and $\cosh x = \cos ix$. Analogously, we can define *modified Bessel functions* (or *hyperbolic Bessel functions*) of the first and second kind as follows:

$$I_n(x) = i^{-n} J_n(ix) \tag{13.54}$$

and

$$K_n(x) = \frac{\pi}{2} i^{n+1} \big[J_n(ix) + i Y_n(ix) \big] = \frac{\pi}{2} i^{n+1} H_n^{(1)}(ix). \tag{13.55}$$

Their asymptotic forms for large x are

$$I_n(x) \approx \frac{e^x}{\sqrt{2\pi x}} \quad \text{and} \quad K_n(x) \approx \sqrt{\frac{\pi}{2x}} e^{-x} \quad \text{for} \quad x \gg n. \tag{13.56}$$

We have given just a meager sampling of formulas involving Bessel functions. Many more can be found in Abramowitz & Stegun and other references. G.N. Whittaker's classic *A Treatise on the Theory of Bessel Functions* (Cambridge University Press, 1952) is a ponderous volume devoted entirely to the subject.

13.7 Laplace's Equation in Spherical Coordinates

Laplace's equation in spherical polar coordinates can be written

$$\nabla^2 \Psi(r, \theta, \phi) = \frac{1}{r^2} \frac{\partial}{\partial r} r^2 \frac{\partial \Psi}{\partial r} + \frac{1}{r^2 \sin\theta} \frac{\partial}{\partial \theta} \sin\theta \frac{\partial \Psi}{\partial \theta} + \frac{1}{r^2 \sin^2\theta} \frac{\partial^2 \Psi}{\partial \phi^2} = 0. \tag{13.57}$$

We consider separable solutions

$$\Psi(r, \theta, \phi) = R(r) Y(\theta, \phi), \tag{13.58}$$

where the functions $Y(\theta, \phi)$ are known as a *spherical harmonics*. Substitution of (13.58) into (13.57), followed by division by $R(r)Y(\theta, \phi)$ and multiplication by r^2, separates the radial and angular variables:

$$\frac{1}{Y(\theta, \phi)} \left[\frac{1}{\sin \theta} \frac{\partial}{\partial Y} \sin \theta \frac{\partial Y}{\partial \theta} + \frac{1}{\sin^2 \theta} \frac{\partial^2 Y}{\partial \phi^2} \right] + \frac{1}{R(r)} \left[r^2 R''(r) + 2r R'(r) \right] = 0.$$

(13.59)

This can hold true for all values of r, θ, ϕ only if each of the two parts of this partial differential equation equals a constant. Let the first part equal $-\lambda$ and the second equal λ. The ordinary differential equation for r becomes

$$r^2 R''(r) + 2r R(r) - \lambda R(r) = 0.$$

(13.60)

It is easily verified that the general solution is

$$R(r) = c_1 r^{\ell} + c_2 r^{-\ell-1}, \quad \text{where} \quad \lambda = \ell(\ell + 1).$$

(13.61)

Returning to the spherical harmonics, these evidently satisfy a partial differential equation in two variables:

$$\frac{1}{\sin \theta} \frac{\partial}{\partial \theta} \sin \theta \frac{\partial Y}{\partial \theta} + \frac{1}{\sin^2 \theta} \frac{\partial^2 Y}{\partial \phi^2} + \lambda Y = 0.$$

(13.62)

Again we assume a separable solution with

$$Y(\theta, \phi) = \Theta(\theta)\Phi(\phi).$$

(13.63)

Substituting (13.63) into (13.62), dividing by $\Theta(\theta)\Phi(\phi)$, and multiplying by $\sin^2 \theta$, we achieve separation of variables:

$$\frac{\sin \theta}{\Theta(\theta)} \frac{d}{d\theta} \sin \theta \frac{d\Theta}{d\theta} + \frac{1}{\Phi(\phi)} \frac{d^2\Phi}{d\phi^2} + \lambda \sin^2 \theta = 0.$$

(13.64)

Setting the term containing ϕ to a constant $-m^2$, we obtain the familiar differential equation

$$\Phi''(\phi) + m^2 \Phi(\phi) = 0.$$

(13.65)

Solutions periodic in 2π can be chosen as

$$\Phi_m(\phi) = \sqrt{\frac{1}{2\pi}} e^{im\phi} \quad \text{with} \quad m = 0, \pm 1, \pm 2, \dots$$

(13.66)

Alternative solutions are $\sin m\phi$ and $\cos m\phi$.

13.8   Legendre Polynomials

Separation of variables in Laplace's equation leads to an ODE for the function $\Theta(\theta)$:

$$\left\{ \frac{1}{\sin\theta} \frac{d}{d\theta} \sin\theta \frac{d}{d\theta} - \frac{m^2}{\sin^2\theta} + \lambda \right\} \Theta(\theta) = 0. \tag{13.67}$$

Let us first consider the case $m = 0$ and define a new independent variable

$$x = \cos\theta \quad \text{with} \quad P(x) = \Theta(\theta). \tag{13.68}$$

This transforms Eq. (13.67) to

$$(1 - x^2)P''(x) - 2xP'(x) + \lambda P(x) = 0, \tag{13.69}$$

which is known as *Legendre's differential equation*. We can construct a solution to this linear second-order equation by exploiting Leibniz's formula (13.29). Begin with the function

$$u = (1 - x^2)^\ell, \tag{13.70}$$

which is a solution of the first-order equation

$$(1 - x^2)u'(x) + 2\ell x u(x) = 0. \tag{13.71}$$

Differentiating $(\ell + 1)$ times, we obtain

$$(1 - x^2)p''(x) - 2xp'(x) + \ell(\ell + 1)p(x) = 0, \tag{13.72}$$

where

$$p(x) = \frac{d^\ell u}{dx^\ell} = \frac{d^\ell}{dx^\ell}(1 - x^2)^\ell. \tag{13.73}$$

This is a solution of Eq. (13.69) for

$$\lambda = \ell(\ell + 1), \quad \ell = 0, 1, 2, \ldots \tag{13.74}$$

With a choice of constant such that $P_\ell(1) = 1$, the Legendre polynomials can be defined by *Rodrigues' formula*:

$$P_\ell(x) = \frac{1}{2^\ell \ell!} \frac{d^\ell}{dx^\ell}(1 - x^2)^\ell. \tag{13.75}$$

Reverting to the original variable θ, the first few Legendre polynomials are

$$P_0(\cos\theta) = 1, \quad P_1(\cos\theta) = \cos\theta,$$

$$P_2(\cos\theta) = \frac{1}{2}(3\cos^2\theta - 1),$$

$$P_3(\cos\theta) = \frac{1}{2}(5\cos^3\theta - 3\cos\theta),$$

$$P_4(\cos\theta) = \frac{1}{8}(35\cos^4\theta - 30\cos^2\theta + 3). \tag{13.76}$$

The Legendre polynomials obey the orthonormalization relations

$$\int_{-1}^{+1} P_\ell(x) P_{\ell'}(x) dx = \int_0^\pi P_\ell(\cos\theta) P_{\ell'}(\cos\theta) \sin\theta\, d\theta = \frac{2}{2\ell+1} \delta_{\ell,\ell'}. \quad (13.77)$$

A generating function for Legendre polynomials is given by

$$\left(1 - 2tx + t^2\right)^{-1} = \sum_{\ell=0}^\infty P_\ell(x) t^\ell. \quad (13.78)$$

An alternative generating function involves Bessel function of order zero:

$$e^{tx} J_0\left(t\sqrt{1-x^2}\right) = \sum_{\ell=0}^\infty \frac{P_\ell(x)}{\ell!} t^\ell. \quad (13.79)$$

Another remarkable connection between Legendre polynomials and Bessel functions is the relation

$$\lim_{n\to\infty} P_n\left(\cos\frac{z}{n}\right) = J_0(z). \quad (13.80)$$

Returning to Eq. (13.67) for arbitrary values of m, the analog of Eq. (13.69) can be written

$$(1 - x^2) P''(x) - 2x P'(x) + \left[\ell(\ell+1) - \frac{m^2}{1-x^2}\right] P(x) = 0. \quad (13.81)$$

The solutions are readily found to be

$$P_\ell^m(x) = (1 - x^2)^{|m|/2} \frac{d^{|m|}}{dx^{|m|}} P_\ell(x) \quad (13.82)$$

known as *associated Legendre functions*. These reduce to the Legendre polynomials (13.75) when $m = 0$. Since $P_\ell(x)$ is a polynomial of degree ℓ, $|m|$ is limited to the values $0, 1, 2, \ldots, \ell$. The associated Legendre functions obey the orthonormalization relations

$$\int_{-1}^{+1} P_\ell^m(x) P_{\ell'}^m(x) dx = \int_0^\pi P_\ell^m(\cos\theta) P_{\ell'}^m(\cos\theta) \sin\theta\, d\theta$$

$$= \frac{2}{2\ell+1} \frac{(\ell+|m|)!}{(\ell-|m|)!} \delta_{\ell,\ell'}. \quad (13.83)$$

The orthonormalized solutions to Eq. (13.67) are thus given by

$$\Theta_{\ell m}(\theta) = \left[\frac{2\ell+1}{2} \frac{(\ell-|m|)!}{(\ell+|m|)!}\right]^{1/2} P_\ell^m(\cos\theta). \quad (13.84)$$

13.9 Spherical Harmonics

Combining the above functions of θ and ϕ, we obtain the spherical harmonics

$$Y_{\ell m}(\theta, \phi) = \epsilon_m \Theta_{\ell m}(\theta) \Phi_m(\phi) = \epsilon_m \left[\frac{2\ell + 1}{4\pi} \frac{(\ell - |m|)!}{(\ell + |m|)!} \right]^{1/2} P_\ell^m(\cos\theta) e^{im\phi},$$

(13.85)

where

$$\epsilon_m = \begin{cases} 1 & \text{for } m \leqslant 0, \\ (-1)^m & \text{for } m > 0. \end{cases}$$

(13.86)

The factor ϵ_m is appended in applications to quantum mechanics, in accordance with the Condon and Shortley phase convention. Two important special cases are

$$Y_{\ell 0}(\theta, \phi) = \left[\frac{2\ell + 1}{4\pi} \right]^{1/2} P_\ell(\cos\theta)$$

(13.87)

and

$$Y_{\ell\ell}(\theta, \phi) = \frac{(-)^\ell}{2^\ell \ell!} \left[\frac{(2\ell + 1)!}{4\pi} \right]^{1/2} \sin^\ell \theta \, e^{i\ell\phi}.$$

(13.88)

Following is a listing of spherical harmonics $Y_{\ell m}(\theta, \phi)$ through $\ell = 2$:

$$Y_{00} = \left(\frac{1}{4\pi} \right)^{1/2}, \quad Y_{10} = \left(\frac{3}{4\pi} \right)^{1/2} \cos\theta,$$

$$Y_{1\pm1} = \mp \left(\frac{3}{4\pi} \right)^{1/2} \sin\theta \, e^{\pm i\phi}, \quad Y_{20} = \left(\frac{5}{16\pi} \right)^{1/2} (3\cos^2\theta - 1),$$

$$Y_{2\pm1} = \mp \left(\frac{15}{8\pi} \right)^{1/2} \cos\theta \sin\theta \, e^{\pm i\phi}, \quad Y_{2\pm2} = \left(\frac{15}{32\pi} \right)^{1/2} \sin^2\theta \, e^{\pm 2i\phi}.$$

A graphical representation of these functions is given in Figure 13.2. Surfaces of constant absolute value are drawn, with intermediate shadings representing differing complex values of the functions. In quantum mechanics, spherical harmonics are eigenfunctions of orbital angular momentum operators such that

$$L^2 Y_{\ell m} = \ell(\ell + 1)\hbar^2 Y_{\ell m}$$

(13.89)

and

$$L_z Y_{\ell m} = m\hbar Y_{\ell m}.$$

(13.90)

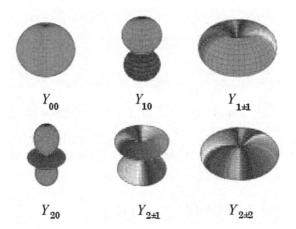

Y_{00} Y_{10} $Y_{1\pm1}$

Y_{20} $Y_{2\pm1}$ $Y_{2\pm2}$

Figure 13.2 Contours of spherical harmonics as three-dimensional polar plots.

The spherical harmonics are an orthonormal set with respect to integration over solid angle:

$$\int_0^\pi \int_0^{2\pi} Y_{\ell'm'}^*(\theta, \phi) Y_{\ell m}(\theta, \phi) \sin\theta \, d\theta \, d\phi = \delta_{\ell\ell'}\delta_{mm'}. \tag{13.91}$$

Linear combinations of $Y_{\ell m}$ and $Y_{\ell -m}$ contain the real functions:

$$P_\ell^m(\cos\theta) \begin{cases} \sin m\phi, \\ \cos m\phi. \end{cases} \tag{13.92}$$

These are called *tesseral harmonics* since they divide the surface of a sphere into *tesserae*—four-sided figures bounded by nodal lines of latitude and longitude. (Nodes are places where the wavefunction equals zero.) When $m = 0$, the spherical harmonics are real functions:

$$Y_{\ell 0}(\theta, \phi) = \left[\frac{2\ell + 1}{4\pi}\right]^{1/2} P_\ell(\cos\theta). \tag{13.93}$$

These correspond to *zonal harmonics* since their nodes are circles of latitude which divide the surface of the sphere into zones. When $m = \ell$, the spherical harmonics reduce to

$$Y_{\ell\ell}(\theta, \phi) = \frac{(-)^\ell}{2^\ell \ell!} \left[\frac{(2\ell + 1)!}{4\pi}\right]^{1/2} \sin^\ell\theta \, e^{i\ell\phi}. \tag{13.94}$$

The corresponding real functions

$$\sin^\ell\theta \begin{cases} \sin \ell\phi, \\ \cos \ell\phi, \end{cases} \tag{13.95}$$

are called *sectoral harmonics*. The three types of surface harmonics are shown in Figure 13.3.

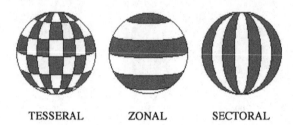

TESSERAL   ZONAL   SECTORAL

Figure 13.3 Three types of spherical harmonics plotted on surface of a sphere. Boundaries between shaded (positive) and white (negative) regions are nodes where wavefunction equals zero.

13.10 Spherical Bessel Functions

Helmholtz's equation in spherical polar coordinates can be obtained by adding a constant to Laplace's equation (13.57):

$$\left(\nabla^2 + k^2\right)\Psi(r, \theta, \phi) = \frac{1}{r^2}\frac{\partial}{\partial r}r^2\frac{\partial\Psi}{\partial r} + \frac{1}{r^2\sin\theta}\frac{\partial}{\partial\theta}\sin\theta\frac{\partial\Psi}{\partial\theta}$$
$$+ \frac{1}{r^2\sin^2\theta}\frac{\partial^2\Psi}{\partial\phi^2} + k^2\Psi = 0. \tag{13.96}$$

As in the case of Laplace's equation, Eq. (13.96) has separable solutions

$$\Psi(r, \theta, \phi) = R(r)Y_{\ell m}(\theta, \phi). \tag{13.97}$$

Since we now know all about spherical harmonics, we can proceed directly to the radial equation:

$$R''(r) + \frac{2}{r}R'(r) - \frac{\ell(\ell+1)}{r^2}R(r) + k^2R(r) = 0. \tag{13.98}$$

Were it not for the factor 2 in the second term, this would have the form of Bessel's equation (13.35). As in Section 13.5, we redefine the variables to $x = kr$ and $f(x) = R(r)$. We also rewrite ℓ as n to conform to conventional notation. The differential equation now reads

$$x^2 f''(x) + 2xf'(x) + \left[x^2 - n(n+1)\right]f(x) = 0. \tag{13.99}$$

With the substitution $f(x) = x^{-1/2}F(x)$, the equation reduces to

$$x^2 F''(x) + xF'(x) + \left[x^2 - \left(n + \frac{1}{2}\right)^2\right]F(x) = 0, \tag{13.100}$$

which is recognized as Bessel's equation of odd-half order $\frac{1}{2}, \frac{3}{2}, \frac{5}{2} \ldots$ (if n is an integer). The linearly independent solutions are $J_{n+1/2}(x)$ and $Y_{n+1/2}(x)$, the latter being proportional to $J_{-n-1/2}(x)$ since the order is not an integer. The solutions to Eq. (13.99) are known as *spherical Bessel functions* and conventionally defined by

$$j_n(x) = \sqrt{\frac{\pi}{2x}} J_{n+1/2}(x) \tag{13.101}$$

and

$$y_n(x) = \sqrt{\frac{\pi}{2x}} Y_{n+1/2}(x) = (-)^{n+1} \sqrt{\frac{\pi}{2x}} J_{-n-1/2}(x). \tag{13.102}$$

You can verify that Eq. (13.99) with $n = 0$ has the simple solutions $\sin x/x$ and $\cos x/x$. In fact, spherical Bessel functions have closed-form expressions in terms of trigonometric functions and powers of x, given by

$$j_n(x) = (-x)^n \left(\frac{1}{x}\frac{d}{dx}\right)^n \frac{\sin x}{x}, \quad y_n(x) = -(-x)^n \left(\frac{1}{x}\frac{d}{dx}\right)^n \frac{\cos x}{x}. \tag{13.103}$$

Explicit formulas for the first few spherical Bessel functions follow:

$$j_0(x) = \frac{\sin x}{x}, \quad j_1(x) = \frac{\sin x}{x^2} - \frac{\cos x}{x}, \quad j_2(x) = \left(\frac{3}{x^2} - 1\right)\frac{\sin x}{x} - \frac{3\cos x}{x^2}$$

$$y_0(x) = -\frac{\cos x}{x}, \quad y_1(x) = -\frac{\cos x}{x^2} - \frac{\sin x}{x}, \quad y_2(x) = \left(-\frac{3}{x^2} + 1\right)\frac{\cos x}{x} - \frac{3\sin x}{x^2}.$$

$$\tag{13.104}$$

There are also spherical analogs of the Hankel functions:

$$h_n^{(1,2)}(x) = \sqrt{\frac{\pi}{2x}} H_{n+1/2}^{(1,2)}(x) = j_n(x) \pm i y_n(x). \tag{13.105}$$

The first few are

$$h_0^{(1,2)}(x) = \mp \frac{e^{\pm ix}}{x}, \quad h_1^{(1,2)}(x) = -\frac{x \pm i}{x^2} e^{\pm ix},$$

$$h_2^{(1,2)}(x) = \pm i \frac{x^2 \pm 3ix - 3}{x^3} e^{\pm ix}. \tag{13.106}$$

13.11 Hermite Polynomials

The quantum-mechanical harmonic oscillator satisfies the Schrödinger equation:

$$-\frac{\hbar^2}{2\mu}\psi''(x) + \frac{1}{2}kx^2\psi(x) = E\psi(x). \tag{13.107}$$

To reduce the problem to its essentials, simplify the constants with $\hbar = \mu = k = 1$, or alternatively, replace $(\mu k/\hbar^2)^{1/4} x$ by x. Correspondingly, $E = \frac{1}{2}\lambda\hbar\sqrt{k/\mu}$. We must now solve a second-order differential equation with nonconstant coefficients:

$$\psi''(x) + (\lambda - x^2)\psi(x) = 0. \tag{13.108}$$

A useful first step is to determine the asymptotic solution to this equation, giving the form of $\psi(x)$ as $x \to \pm\infty$. For sufficiently large values of $|x|$, $x^2 \gg \lambda$, so that the differential equation can be approximated by

$$\psi''(x) - x^2\psi(x) \approx 0. \tag{13.109}$$

This suggests the following manipulation:

$$\left(\frac{d^2}{dx^2} - x^2\right)\psi(x) \approx \left(\frac{d}{dx} - x\right)\left(\frac{d}{dx} + x\right)\psi(x) \approx 0. \tag{13.110}$$

Now, the first-order differential equation

$$\psi'(x) + x\psi(x) \approx 0 \tag{13.111}$$

can be solved exactly to give

$$\psi(x) \approx \mathrm{const}\, e^{-x^2/2} \quad \text{for} \quad |x| \to \infty. \tag{13.112}$$

To build in this asymptotic behavior, let

$$\psi(x) = H(x)e^{-x^2/2}. \tag{13.113}$$

This reduces Eq. (13.108) to a differential equation for $H(x)$:

$$H''(x) - 2xH'(x) + (\lambda - 1)H(x) = 0. \tag{13.114}$$

To construct a solution to Eq. (13.114), we begin with the function

$$u(x) = e^{-x^2}, \tag{13.115}$$

which is clearly the solution of the *first-order* differential equation

$$u'(x) + 2xu(x) = 0. \tag{13.116}$$

Differentiating this equation $(n + 1)$ times using Leibniz's formula (13.29), we obtain

$$w''(x) + 2xw'(x) + 2(n - 1)w(x) = 0, \tag{13.117}$$

where

$$w(x) = \frac{d^n u}{dx^n} = \frac{d^n}{dx^n}e^{-x^2} = H(x)e^{-x^2}. \tag{13.118}$$

We find that $H(x)$ satisfies

$$H''(x) - 2xH'(x) + 2nH(x) = 0, \tag{13.119}$$

which is known as *Hermite's differential equation*. The solutions in the form

$$H_n(x) = (-)^n e^{x^2}\frac{d^n}{dx^n}e^{-x^2} \tag{13.120}$$

are known as *Hermite polynomials*, the first few of which are enumerated below:

$$H_0(x) = 1, \quad H_1(x) = 2x, \quad H_2(x) = 4x^2 - 2,$$
$$H_3(x) = 8x^3 - 12x, \quad H_4(x) = 16x^4 - 48x^2 + 12. \tag{13.121}$$

Comparing Eq. (13.119) with Eq. (13.114), we can relate the parameters

$$\lambda - 1 = 2n. \tag{13.122}$$

Referring back to the original harmonic-oscillator equation (13.107), this leads to the general formula for energy eigenvalues

$$E_n = \frac{1}{2}\hbar\omega\lambda = \left(n + \frac{1}{2}\right)\hbar\omega \quad \text{with} \quad \omega = \sqrt{\frac{k}{\mu}}. \tag{13.123}$$

A generating function for Hermite polynomials is given by

$$e^{x^2-(t-x)^2} = e^{2tx-t^2} = \sum_{n=0}^{\infty} \frac{H_n(x)}{n!} t^n. \tag{13.124}$$

Using the generating function, we can evaluate integrals over products of Hermite polynomials, such as

$$\int_{-\infty}^{\infty} H_n(x)H_{n'}(x)e^{-x^2}\,dx = 2^n n!\sqrt{\pi}\,\delta_{n,n'}. \tag{13.125}$$

Thus, the functions

$$\psi_n(x) = (2^n n!\sqrt{\pi})^{-1/2} H_n(x)e^{-x^2/2}, \quad n = 0, 1, 2, \dots \tag{13.126}$$

form an orthonormal set with

$$\int_{-\infty}^{\infty} \psi_n(x)\psi_{n'}(x)\,dx = \delta_{n,n'}. \tag{13.127}$$

13.12 Laguerre Polynomials

The quantum-mechanical problem of a particle moving in a central field is represented by a three-dimensional Schrödinger equation with a spherically symmetric potential $V(r)$:

$$\left\{-\frac{\hbar^2}{2m}\nabla^2 + V(r)\right\}\psi(r,\theta,\phi) = E\psi(r,\theta,\phi). \tag{13.128}$$

As in the case of Helmholtz's equation, we have separability in spherical polar coordinates: $\psi(r, \theta, \phi) = R(r)Y_{\ell m}(\theta, \phi)$. In convenient units with $\hbar = m = 1$, the ODE for the radial function can be written

$$-\frac{1}{2}\left[R''(r) + \frac{2}{r}R'(r)\right] + \left[\frac{\ell(\ell+1)}{2r^2} + V(r) - E\right]R(r) = 0. \tag{13.129}$$

We consider the electron in a hydrogen atom or hydrogenlike ion ($He^+, Li^{2+}, \ldots$) "orbiting" around a nucleus of atomic number Z. The attractive Coulomb potential in atomic units ($e^2/4\pi\epsilon_0 \equiv 1$) can be written

$$V(r) = -\frac{Z}{r}.$$
(13.130)

It is again useful to find asymptotic solutions to the differential equation. When $r \to \infty$ the equation is approximated by

$$R''(r) - 2|E|R(r) \approx 0$$
(13.131)

noting that the energy E will be negative for bound states of the hydrogenlike system. We find the asymptotic solution

$$R(r) \approx e^{-\sqrt{2|E|}r} \quad \text{as} \quad r \to \infty.$$
(13.132)

As $r \to 0$, Eq. (13.129) is approximated by

$$R''(r) + \frac{2}{r}R'(r) - \frac{\ell(\ell+1)}{r^2}R(r) \approx 0,$$
(13.133)

which is just Laplace's equation in spherical coordinates. The solution finite at $r = 0$ suggests the limiting dependence

$$R(r) \approx r^\ell \quad \text{as} \quad r \to 0.$$
(13.134)

We can incorporate both limiting forms by writing

$$R(r) = \rho^\ell e^{-\rho/2}L(\rho)$$
(13.135)

in terms of a new variable

$$\rho = 2Zr/n \quad \text{with} \quad E = -Z^2/2n^2,$$
(13.136)

where n is a constant to be determined. The differential equation for $L(\rho)$ then works out to

$$\rho L''(\rho) + (2\ell + 2 - \rho)L'(\rho) + [n - (\ell+1)]L(\rho) = 0.$$
(13.137)

Following the strategy used to solve the Hermite and Legendre differential equations, we begin with a function

$$u(x) = x^\alpha e^{-x},$$
(13.138)

where α is a positive integer. This satisfies the first-order differential equation

$$xu'(x) + (x - \alpha)u(x) = 0.$$
(13.139)

Differentiating this equation ($\alpha + 1$) times using Leibniz's formula (13.29), we obtain

$$xw''(x) + (1 - x)w'(x) + \alpha w(x) = 0,$$
(13.140)

where

$$w(x) = \frac{d^\alpha}{dx^\alpha}(x^\alpha e^{-x}) = e^{-x}L(x). \tag{13.141}$$

Laguerre polynomials are defined by *Rodrigues' formula*:

$$L_\alpha(x) = \frac{e^x}{\alpha!}\frac{d^n}{dx^\alpha}(x^\alpha e^{-x}). \tag{13.142}$$

We require a generalization known as *associated Laguerre polynomials*, defined by

$$L_\alpha^\beta(x) = (-)^\beta \frac{d^\beta}{dx^\beta}L_{\alpha+\beta}(x). \tag{13.143}$$

These are solutions of the differential equation

$$xL''(x) + (\beta + 1 - x)L'(x) + (\alpha - \beta)L(x) = 0. \tag{13.144}$$

Comparing Eqs. (13.137) and (13.144), we can identify

$$\beta = 2\ell + 1, \quad \alpha = n + \ell, \tag{13.145}$$

where n must be a positive integer. The bound-state energy hydrogenlike eigenvalues are therefore determined:

$$E_n = -\frac{Z^2}{2n^2} \text{ atomic units}, \quad n = 1, 2, \dots \tag{13.146}$$

with the normalized radial functions

$$R_{n\ell}(r) = N_{n\ell}\rho^\ell L_{n+\ell}^{2\ell+1}(\rho)e^{-\rho/2}, \quad \rho = 2Zr/n. \tag{13.147}$$

The conventional definition of the constant is

$$N_{n\ell} = -\left[\frac{(n-\ell-1)!}{2n\{(n-\ell)!\}^3}\right]^{1/2}\left(\frac{2Z}{n}\right)^{3/2} \tag{13.148}$$

such that

$$\int_0^\infty \left[R_{n\ell}(r)\right]^2 r^2\, dr = 1. \tag{13.149}$$

Laguerre and associated Laguerre polynomials can be found from the following generating functions:

$$(1-t)^{-1}\exp\left(-\frac{xt}{1-t}\right) = \sum_{n=0}^\infty \frac{L_n(x)}{n!}t^n, \tag{13.150}$$

$$(1-t)^{-1}\left(\frac{-t}{1-t}\right)^k \exp\left(-\frac{xt}{1-t}\right) = \sum_{n=k}^\infty \frac{L_n^k(x)}{n!}t^n. \tag{13.151}$$

13.13 Hypergeometric Functions

A *geometric series* is a function of whose terms constitute a geometric progression, in which the ratio of successive terms is equal. Consider, for example, with z and a, in general, being complex quantities,

$$F(z) = 1 + az + a^2z^2 + a^3z^3 + \cdots = \sum_{n=0}^{\infty} a^n z^n. \tag{13.152}$$

This series converges to the value $1/(1-az)$, provided that $|az| < 1$. A certain type of generalization of a geometric series is known as a *hypergeometric series* or *hypergeometric function*. This has the form of a power series in z in which the coefficients a^n are replaced by ratios of rational functions of constants.

A rudimentary example of a hypergeometric function can be written

$$_1F_0(a; \_; z) = 1 + a\frac{z}{1!} + a(a+1)\frac{z^2}{2!} + a(a+1)(a+2)\frac{z^3}{3!} + \cdots = \sum_{n=0}^{\infty}(a)_n\frac{z^n}{n!}. \tag{13.153}$$

Here $(a)_n$, known as the *Pochhammer symbol* (also called a rising or ascending factorial), is defined by

$$(a)_0 = 1, \ (a)_1 = a, \ (a)_2 = a(a+1), (a)_3 = a(a+1)(a+2), \ldots,$$
$$(a)_n = a(a+1)(a+2)\cdots(a+n-1) \tag{13.154}$$

or

$$(a)_n = \frac{(a+n-1)!}{(a-1)!} = \frac{\Gamma(a+n)}{\Gamma(a)}. \tag{13.155}$$

The last form is valid even if a is not an integer. For $a = 1$, $_1F_0$ reduces to the geometric series for z. When a is a negative integer, the series terminates, for example,

$$_1F_0(-2; \_; z) = 1 + (-2)z + \frac{(-2)(-2+1)}{2!}z^2 + 0. \tag{13.156}$$

Problem 13.13.1. Show that the Pochhammer symbol $(1)_n = n!$.

Problem 13.13.2. Show that $_1F_0(a; \_; z)$ is a solution of the first-order differential equation

$$(1-z)y'(z) - ay(z) = 0 \tag{13.157}$$

and that, in fact, $_1F_0(a; \_; z) = (1-z)^{-a}$.

Next consider the hypergeometric function

$$_1F_1(a; c; z) = 1 + \frac{a}{c}\frac{z}{1!} + \frac{a(a+1)}{c(c+1)}\frac{z^2}{2!} + \frac{a(a+1)(a+2)}{c(c+1)(c+2)}\frac{z^3}{3!} + \cdots = \sum_{n=0}^{\infty} \frac{(a)_n}{(c)_n} \frac{z^n}{n!}.$$

(13.158)

The series is convergent for finite z, provided that $c \neq 0, -1, -2, \ldots$ It is known as the *confluent hypergeometric function* (explanation later) or *Kummer function*. It is a solution of the differential equation:

$$zy''(z) + (c - z)y'(z) - ay = 0.$$

(13.159)

Two important properties of the confluent geometric function are Kummer's first formula, again for $c \neq 0$ or a negative integer,

$$_1F_1(a; c; z) = e^z {}_1F_1(c - a; c; -z)$$

(13.160)

and Kummer's second formula:

$$_1F_1(a; 2a; z) = e^z {}_1F_1(a; 2a; -z).$$

(13.161)

For large $|z|$, the function has the asymptotic forms:

$$_1F_1(a; c; z) \approx \begin{cases} \frac{\Gamma(c)}{\Gamma(a)} e^z z^{a-c} & \text{for } \mathcal{R}z > 0, \\ \frac{\Gamma(c)}{\Gamma(c-a)} (-z)^{-a} & \text{for } \mathcal{R}z < 0. \end{cases}$$

(13.162)

For certain combinations of a, c, and z, the confluent hypergeometric function reduces to some of the special functions considered earlier in this chapter. First, the rather trivial case:

$$_1F_1(a; a; z) = e^z.$$

(13.163)

(Even more trivial is $_0F_0(\_; \_; z) = e^z$.) Some other elementary functions are obtained from

$$_1F_1(1; 2; -2ix) = \frac{e^{-ix}}{x} \sin x$$

(13.164)

and

$$_1F_1(1; 2; 2x) = \frac{e^x}{x} \sinh x.$$

(13.165)

The error function can be represented by

$$\text{erf}(x) = \frac{2x}{\sqrt{\pi}} {}_1F_1\left(\frac{1}{2}; \frac{3}{2}; -x^2\right).$$

(13.166)

When $c = 2a$, we obtain a relation for Bessel functions:

$$J_\nu(z) = \frac{1}{\Gamma(\nu+1)} \left(\frac{z}{2}\right)^\nu e^{-iz} {}_1F_1\left(\nu+\frac{1}{2}; 2\nu+1; 2iz\right).$$ (13.167)

Another formula for a Bessel function:

$$J_\nu(z) = \frac{1}{\Gamma(\nu+1)} \left(\frac{z}{2}\right)^\nu {}_0F_1(\_; \nu+1; -z^2/4).$$ (13.168)

For $n = 0, 1, 2, \ldots$, the Laguerre polynomial is given simply by

$$L_n(x) = {}_1F_1(-n; 1; x).$$ (13.169)

Hermite polynomials reduce slightly to different forms for even and odd n. For $n = 0, 2, 4, \ldots$,

$$H_n(x) = (-2)^{n/2}(n-1)!\, {}_1F_1\left(-\frac{n}{2}; \frac{1}{2}; x^2\right),$$ (13.170)

while, for $n = 1, 3, 5, \ldots$,

$$H_n(x) = -(-2)^{(n-1)/2} n!\, {}_1F_1\left(\frac{1-n}{2}; \frac{3}{2}; x^2\right).$$ (13.171)

Gauss' famous hypergeometric differential equation is given by

$$z(1-z)y''(z) + [c - (a+b+1)z]y'(z) - aby(z) = 0,$$ (13.172)

where c does not equal $0, -1, -2, \ldots$ There exist 24 possible solutions involving transformed versions of ${}_2F_1$. We will consider only the simplest one, which is what was originally called *the* hypergeometric function:

$$ {}_2F_1(a, b; c; z) = 1 + \frac{ab}{c}\frac{x}{1!} + \frac{a(a+1)b(b+1)}{c(c+1)}\frac{z^2}{2!} + \cdots = \sum_{n=0}^\infty \frac{(a)_n(b)_n}{(c)_n}\frac{z^n}{n!}.$$ (13.173)

There is a huge amount of mathematical technology involving the hypergeometric function. The interested reader can consult numerous references which cover these in great detail. We will content ourselves with some simple results:

$$ {}_2F_1(1, 1; 2; -x) = \frac{\ln(1+x)}{x},$$ (13.174)

$$ {}_2F_1\left(\frac{1}{2}, \frac{1}{2}; \frac{3}{2}; x^2\right) = \frac{\arcsin x}{x},$$ (13.175)

$$ {}_2F_1\left(\frac{1}{2}, 1; \frac{3}{2}; -x^2\right) = \frac{\arctan x}{x}.$$ (13.176)

The Legendre polynomials can be represented by:

$$P_n(x) = {}_2F_1\left(-n, n+1; 1; \frac{1-x}{2}\right).$$ (13.177)

The concept of *confluence* involves the substitution $z \to z/b$, followed by taking the limit $b \to \infty$. This reduces the hypergeometric differential equation and the hypergeometric function to their analogs for the *confluent hypergeometric function*.

It is also possible to define generalized hypergeometric functions of higher order, involving more than three constants:

$$ {}_pF_q(a_1, a_2, \ldots, a_p; c_1, c_2, \ldots, c_q; z) = \sum_{n=0}^{\infty} \frac{(a_1)_n(a_2)_n \cdots (a_p)_n}{(c_1)_n(c_2)_n \cdots (c_q)_n} \frac{z^n}{n!}. $$ (13.178)

14 Complex Variables

A deeper understanding of functional analysis, even principles involving real functions of real variables, can be attained if the functions and variables are extended into the complex plane. Figure 14.1 shows schematically how a functional relationship $w = f(z)$ can be represented by a mapping of the z-plane, with $z = x + iy$, into the w-plane, with $w = u + iv$.

14.1 Analytic Functions

Let

$$w(x, y) = u(x, y) + iv(x, y) \tag{14.1}$$

be a complex-valued function constructed from two real functions $u(x, y)$ and $v(x, y)$. Under what conditions can $w(x, y)$ be considered a legitimate function of a single *complex variable* $z = x + iy$, allowing us to write $w = w(z)$? A simple example would be

$$u(x, y) = x^2 - y^2, \quad v(x, y) = 2xy, \tag{14.2}$$

so that

$$w(x, y) = (x^2 - y^2) + i(2xy) = (x + iy)^2. \tag{14.3}$$

Evidently

$$w(z) = z^2 \quad \text{with } \Re w = x^2 - y^2, \ \Im w = 2xy. \tag{14.4}$$

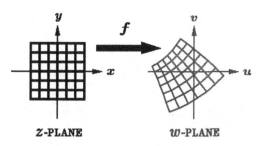

Figure 14.1 Mapping of the functional relation $w = f(z)$. If the function is analytic, then the mapping is *conformal*, with orthogonality of grid lines preserved.

Guide to Essential Math 2e. http://dx.doi.org/10.1016/B978-0-12-407163-6.00014-X

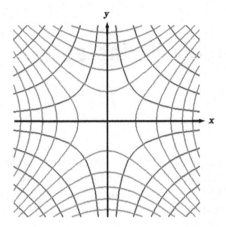

Figure 14.2 Contours of $w(z) = z^2$ in the complex plane: $\Re(z^2) = x^2 - y^2$, $\Im(z^2) = 2xy$.

This function can be represented in the complex plane as shown in Figure 14.2. A counterexample, which is *not* a legitimate function of a complex variable, would be

$$w(x, y) = x^2 + y^2 = zz^* \neq w(z), \tag{14.5}$$

since the complex conjugate $z^* = x - iy$ is *not* considered a function of z. To derive a general condition for $w(x, y) = w(z)$, express x and y in terms of z and z^* using

$$x = (z + z^*)/2, \quad y = (z - z^*)/2i. \tag{14.6}$$

An arbitrary function in Eq. (14.1) can thus be reexpressed in the functional form $w = w(z, z^*)$. The condition that $w = w(z)$, with no dependence on z^*, implies that

$$\frac{\partial w}{\partial z^*} = 0. \tag{14.7}$$

We can write

$$\frac{\partial w}{\partial z^*} = \frac{\partial u}{\partial z^*} + i\frac{\partial v}{\partial z^*} = \frac{\partial u}{\partial x}\frac{\partial x}{\partial z^*} + \frac{\partial u}{\partial y}\frac{\partial y}{\partial z^*} + i\left[\frac{\partial v}{\partial x}\frac{\partial x}{\partial z^*} + \frac{\partial v}{\partial y}\frac{\partial y}{\partial z^*}\right] = 0. \tag{14.8}$$

Using (14.6), this reduces to

$$\frac{1}{2}\frac{\partial u}{\partial x} + \frac{i}{2}\frac{\partial u}{\partial y} + i\left[\frac{1}{2}\frac{\partial v}{\partial x} + \frac{i}{2}\frac{\partial v}{\partial y}\right] = \frac{1}{2}\left(\frac{\partial u}{\partial x} - \frac{\partial v}{\partial y}\right) + \frac{i}{2}\left(\frac{\partial u}{\partial y} + \frac{\partial v}{\partial x}\right) = 0. \tag{14.9}$$

Since the real and imaginary parts must individually equal zero, we obtain the *Cauchy-Riemann equations*:

$$\frac{\partial u}{\partial x} = \frac{\partial v}{\partial y} \quad \text{and} \quad \frac{\partial u}{\partial y} = -\frac{\partial v}{\partial x}. \tag{14.10}$$

These conditions on the real and imaginary parts of a function $w(x, y)$ must be fulfilled in order for w to a function of the complex variable z. If, in addition, u and v have continuous partial derivatives with respect to x and y in some region, then $w(z)$ in that region is an *analytic function* of z. In complex analysis, the term *holomorphic function* is often used to distinguish it from a *real* analytic function.

A complex variable z can alternatively be expressed in polar form

$$z = \rho\, e^{i\theta}, \tag{14.11}$$

where ρ is referred to as the *modulus* and θ, the *phase* or *argument*. Correspondingly, the function $w(z)$ would be written

$$w(\rho, \theta) = u(\rho, \theta) + i v(\rho, \theta). \tag{14.12}$$

The Cauchy-Riemann equations in polar form are then given by

$$\frac{\partial u}{\partial \rho} = \frac{1}{\rho}\frac{\partial v}{\partial \theta} \quad \text{and} \quad \frac{\partial v}{\partial \rho} = -\frac{1}{\rho}\frac{\partial u}{\partial \theta}. \tag{14.13}$$

Consider the function

$$w(z) = \frac{1}{z} = \frac{1}{x + iy} = \frac{x}{x^2 + y^2} - i\frac{y}{x^2 + y^2}. \tag{14.14}$$

Both $u(x, y)$ and $v(x, y)$ and their x- and y-derivatives are well behaved everywhere in the x, y-plane *except* at the point $x = y = 0$, where they become discontinuous and, in fact, infinite. In mathematical jargon, the function and its derivatives *do not exist* at that point. We say therefore that $w(z) = 1/z$ is an analytic function in the entire complex plane *except* at the point $z = 0$. A value of z at which a function is *not* analytic is called a *singular point* or *singularity*.

Taking the x-derivative of the first Cauchy-Riemann equation and the y-derivative of the second, we have

$$\frac{\partial^2 u}{\partial x^2} = \frac{\partial^2 v}{\partial x \partial y} \quad \frac{\partial^2 u}{\partial y^2} = -\frac{\partial^2 v}{\partial y \partial x}. \tag{14.15}$$

Since the mixed second derivatives of $v(x, y)$ are equal,

$$\frac{\partial^2 u}{\partial x^2} + \frac{\partial^2 u}{\partial y^2} = \nabla^2 u(x, y) = 0. \tag{14.16}$$

Analogously, we find

$$\frac{\partial^2 v}{\partial x^2} + \frac{\partial^2 v}{\partial y^2} = \nabla^2 v(x, y) = 0. \tag{14.17}$$

Therefore both the real and imaginary parts of an analytic function are solution of the two-dimensional Laplace equation, known as *harmonic functions*. This can be verified for $u(x, y) = x^2 - y^2$ and $v(x, y) = 2xy$, given in the above example of the analytic function $w(z) = z^2$.

14.2 Derivative of an Analytic Function

The derivative of a complex function is given by the obvious transcription of the definition used for real functions:

$$\frac{df}{dz} = f'(z) \equiv \lim_{\Delta z \to 0} \left[\frac{f(z + \Delta z) - f(z)}{\Delta z} \right]. \tag{14.18}$$

In the definition of a real derivative, such as $\partial f/\partial x$ or $\partial f/\partial y$, there is only one way for Δx or Δy to approach zero. For $\Delta z = \Delta x + i\Delta y$ in the complex plane, there are however an infinite number of ways to approach $\Delta z = 0$. For an analytic function, all of them should give the same result for the derivative.

Let us consider two alternative ways to achieve the limit $\Delta z \to 0$: (1) along the x-axis with $\Delta z = \Delta x$, $\Delta y = 0$ or (2) along the y-axis with $\Delta z = i\Delta y$, $\Delta x = 0$. With $f(z) = u(x, y) + iv(x, y)$, we can write

$$\Delta f = f(z + \Delta z) - f(z) = \Delta u + i\Delta v \tag{14.19}$$

with

$$\Delta u = u(x+\Delta x, y+\Delta y) - u(x, y), \quad \Delta v = v(x+\Delta x, y+\Delta y) - v(x, y). \tag{14.20}$$

The limits for $f'(z)$ in the alternative processes are then given by

$$f_1'(z) = \lim_{\Delta x \to 0} \left[\frac{\Delta u + i\Delta v}{\Delta x} \right] = \frac{\partial u}{\partial x} + i\frac{\partial v}{\partial x} \tag{14.21}$$

and

$$f_2'(z) = \lim_{\Delta y \to 0} \left[\frac{\Delta u + i\Delta v}{i\Delta y} \right] = -i\frac{\partial u}{\partial y} + \frac{\partial v}{\partial y}. \tag{14.22}$$

Equating the real and imaginary parts of (14.21) and (14.22), we again arrive at the Cauchy-Riemann Eqs. (14.10).

All the familiar formulas for derivatives remain valid for complex variables, for example, $d(z^n)/dz = nz^{n-1}$, and so forth.

14.3 Contour Integrals

The integral of a complex function $\int_C f(z)dz$ has the form of a line integral (see Section 11.7) over a specified path or *contour* C between two points z_a and z_b in the complex plane. It is defined as the analogous limit of a Riemann sum:

$$\int_C f(z)dz \equiv \lim_{\substack{n \to \infty \\ \Delta z_i \to 0}} \sum_{i=1}^{n} f(z_i)\Delta z_i, \quad \text{where } \Delta z_i \equiv z_i - z_{i-1}, \tag{14.23}$$

where the points $z_i = x_i + iy_i$ lie on a continuous path C between $z_a = z_0$ and $z_b = z_n$. In the most general case, the value of the integral depends on the path C. For the case of an analytic function in a simply connected region, we will show that the contour integral is *independent* of path, being determined entirely by the endpoints z_a and z_b.

14.4   Cauchy's Theorem

This is the central result in the theory of complex variables. It states that the line integral of an analytic function around an arbitrary closed path in a simple-connected region vanishes:

$$\oint f(z)dz = 0. \tag{14.24}$$

The path of integration is understood to be traversed in the *counterclockwise* sense. An "informal" proof can be based on the identification of $f(z)dz$ with an exact differential expression (see Section 11.6):

$$f(z)dz = [u(x, y) + iv(x, y)](dx + idy)$$
$$= [u(x, y) + iv(x, y)]dx + [iu(x, y) - v(x, y)]dy. \tag{14.25}$$

It is seen that Euler's reciprocity relation (11.48)

$$\frac{\partial u}{\partial y} + i\frac{\partial v}{\partial y} = i\frac{\partial u}{\partial x} - \frac{\partial v}{\partial x} \tag{14.26}$$

is equivalent to the Cauchy-Riemann Eqs. (14.10). Cauchy's theorem is then a simple transcription of the result (11.73) for the line integral around a closed path. The region in play must be simply connected, with no singularities. Equation (14.24) is sometimes referred to as the *Cauchy-Goursat theorem*. Goursat proved it under somewhat less restrictive conditions, showing that $f'(x)$ need not be a continuous function.

14.5   Cauchy's Integral Formula

The most important applications of Cauchy's theorem involve functions with singular points. Consider the integral

$$\oint_C \frac{f(z)}{z - z_0}dz$$

around the closed path C shown in Figure 14.3. Let $f(z)$ be an analytic function in the entire region. Then $f(z)/(z - z_0)$ is also analytic *except* at the point $z = z_0$. The contour C can be shrunk to a small circle C_0 surrounding z_0, as shown in the figure. The infinitesimally narrow channel connecting C to C_0 is traversed in both directions, thus canceling its contribution to the integral around the composite contour. By Cauchy's theorem

$$\oint_C \frac{f(z)}{z - z_0}dz - \oint_{C_0} \frac{f(z)}{z - z_0}dz = 0. \tag{14.27}$$

The minus sign appears because the integration is *clockwise* around the circle C_0. We find therefore

$$\oint_C \frac{f(z)}{z - z_0}dz = \oint_{C_0} \frac{f(z)}{z - z_0}dz = f(z_0)\oint_{C_0} \frac{dz}{z - z_0}, \tag{14.28}$$

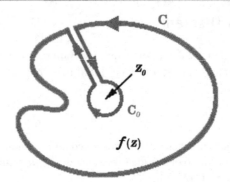

Figure 14.3 Contours for derivation of Cauchy's integral formula: $f(z_0) = \frac{1}{2\pi i} \oint_C \frac{f(z)}{z-z_0} dz$.

assuming that C_0 is a sufficiently small circle that $f(z)$ is nearly constant within, well approximated as $f(z_0)$. It is convenient now to switch to a polar representation of the complex variable, with

$$z - z_0 = \rho e^{i\theta}, \quad dz = i\rho e^{i\theta} d\theta. \tag{14.29}$$

We find then

$$\oint_C \frac{f(z)}{z-z_0} dz = f(z_0) \int_0^{2\pi} \frac{i\rho e^{i\theta} d\theta}{\rho e^{i\theta}} = f(z_0)i \int_0^{2\pi} d\theta = 2\pi i f(z_0). \tag{14.30}$$

The result is *Cauchy's integral theorem*:

$$f(z_0) = \frac{1}{2\pi i} \oint_C \frac{f(z)}{z-z_0} dz. \tag{14.31}$$

A remarkable implication of this formula is a sort of holographic principle. If the values of an analytic function $f(z)$ are known on the boundary of a region, then the value of the function can be determined at every point z_0 *inside* that region.

Cauchy's integral formula can be differentiated with respect to z_0 any number of times to give

$$f'(z_0) = \frac{1}{2\pi i} \oint_C \frac{f(z)}{(z-z_0)^2} dz \tag{14.32}$$

and, more generally,

$$f^{(n)}(z_0) = \frac{n!}{2\pi i} \oint_C \frac{f(z)}{(z-z_0)^{n+1}} dz. \tag{14.33}$$

This shows, incidentally, that derivatives of all orders exist for an analytic function.

14.6 Taylor Series

Taylor's theorem can be derived from the Cauchy integral theorem. Let us first rewrite (14.31) as

$$f(z) = \frac{1}{2\pi i} \oint_C \frac{f(\zeta)}{\zeta - z} d\zeta, \tag{14.34}$$

where ζ is now the variable of integration along the contour C and z, any point in the interior of the contour. Let us develop a power-series expansion of $f(z)$ around the point z_0, also within the contour. Applying the binomial theorem, we can write

$$\frac{1}{\zeta - z} = \frac{1}{(\zeta - z_0) - (z - z_0)} = \frac{1}{\zeta - z_0} \left[1 - \frac{z - z_0}{\zeta - z_0} \right]^{-1} = \sum_{n=0}^{\infty} \frac{(z - z_0)^n}{(\zeta - z_0)^{n+1}}. \tag{14.35}$$

Note that $\zeta > z$ so that $(z - z_0)/(\zeta - z_0) < 1$ and the series converges. Substituting the summation into (14.34), we obtain

$$f(z) = \frac{1}{2\pi i} \sum_{n=0}^{\infty} (z - z_0)^n \oint_C \frac{f(\zeta)}{(\zeta - z_0)^{n+1}} d\zeta. \tag{14.36}$$

Therefore, using Cauchy's integral theorem (14.33),

$$f(z) = \sum_{n=0}^{\infty} \frac{(z - z_0)^n}{n!} f^{(n)}(z_0) = f(z_0) + (z - z_0) f'(z_0) + \frac{(z - z_0)^2}{2} f''(z_0) + \cdots \tag{14.37}$$

This shows that a function analytic in a region can be expanded in a Taylor series about a point $z = z_0$ within that region. The series (14.37) will converge to $f(z)$ within a certain *radius of convergence*, a circle of radius $R < |z_1 - z_0|$, equal to the distance to z_1, the singular point closest to z_0.

We can now understand the puzzling behavior of the series

$$(1 + x)^{-1} = \sum_{n=0}^{\infty} (-1)^n x^n, \tag{14.38}$$

which we encountered in Eq. (7.37). The complex function $f(z) = (1 + z)^{-1}$ has a singularity at $z = -1$. Thus an expansion about $z = 0$ will be valid only within a circle of radius of 1 around the origin. This means that a Taylor series about $z = 0$ will be valid only for $z < 1$. On the real axis this corresponds to $|x| < 1$ and means that both the series $(1 + x)^{-1}$ and $(1 - x)^{-1}$ will converge only under this condition. The function $f(z) = (1 + z)^{-1}$ could, however, be expanded about $z_0 = 1$, giving a larger radius of convergence $|z_0 - (-1)| = 2$. Along the real axis, we find

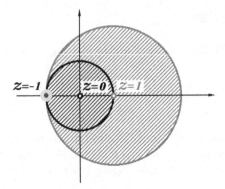

Figure 14.4 Analytic continuation of series expansion for $f(z) = (1 + z)^{-1}$ with singularity at $z = -1$. Expansion about $z = 0$ converges inside small circle, while expansion about $z = 1$ converges inside large circle.

$$f(x) = (1+x)^{-1} = \left[2\left(1 + \frac{x-1}{2}\right)\right]^{-1} = \frac{1}{2}\sum_{n=1}^{\infty}(-)^n\left(\frac{x-1}{2}\right)^n, \qquad (14.39)$$

which converges for $-2 < x < 3$.

The process of shifting the domain of a Taylor series is known as *analytic continuation*. Figure 14.4 shows the circles of convergence for $f(z)$ expanded about $z = 0$ and about $z = 1$. Successive applications of analytic continuation can cover the entire complex plane, exclusive of singular points (with some limitations for multivalued functions).

14.7 Laurent Expansions

Taylor series are valid expansions of $f(z)$ about points z_0 (sometimes called *regular points*) within the region where the function is analytic. It is also possible to expand a function about singular points. Figure 14.5 outlines an annular (shaped like a lock washer) region around a singularity z_0 of a function $f(z)$, but avoiding other singularities at z_1 and z_2. The function is integrated around the contour including C_2 in a counterclockwise sense, C_1 in a clockwise sense, and the connecting cut in canceling directions. Denoting the complex variable on the contour by ζ, we can apply Cauchy's theorem to obtain

$$f(z) = \frac{1}{2\pi i}\oint_C \frac{f(\zeta)}{\zeta - z}d\zeta = \frac{1}{2\pi i}\oint_{C_2}\frac{f(\zeta)}{\zeta - z}d\zeta - \frac{1}{2\pi i}\oint_{C_1}\frac{f(\zeta)}{\zeta - z}d\zeta = 0,$$

$$(14.40)$$

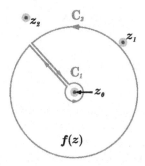

Figure 14.5 Contour for derivation of Laurent expansion of $f(z)$ about singular point $z = z_0$. The singularities at z_1 and z_2 are avoided.

where z is any point within the annular region. On the contour C_2 we have $\zeta > z$ so that $(z - z_0)/(\zeta - z_0) < 1$, validating the convergent expansion (14.35):

$$\frac{1}{\zeta - z} = \sum_{n=0}^{\infty} \frac{(z - z_0)^n}{(\zeta - z_0)^{n+1}}. \tag{14.41}$$

On the contour C_1, however, $z > \zeta$ so that $(\zeta - z_0)/(z - z_0) < 1$ and we have instead

$$\frac{1}{\zeta - z} = -\frac{1}{z - \zeta} = -\sum_{n=0}^{\infty} \frac{(\zeta - z_0)^n}{(z - z_0)^{n+1}} = -\sum_{n=1}^{\infty} (z - z_0)^{-n} (\zeta - z_0)^{(n-1)}, \tag{14.42}$$

where we have inverted the fractions in the last summation and shifted the dummy index. Substituting the last two expansions into (14.40), we obtain

$$f(z) = \frac{1}{2\pi i} \sum_{n=0}^{\infty} (z - z_0)^n \oint_{C_2} \frac{f(\zeta)}{\zeta - z_0} d\zeta$$

$$+ \frac{1}{2\pi i} \sum_{n=1}^{\infty} (z - z_0)^{-n} \oint_{C_1} (\zeta - z_0)^{(n-1)} f(\zeta) d\zeta. \tag{14.43}$$

This is a summation containing both positive and negative powers of $(z - z_0)$:

$$f(z) = \sum_{n=0}^{\infty} a_n (z - z_0)^n + \sum_{n=1}^{\infty} \frac{b_n}{(z - z_0)^n} \tag{14.44}$$

known as a *Laurent series*. The coefficients are given by

$$a_n = \frac{1}{2\pi i} \oint_C \frac{f(z)}{(z - z_0)^{n+1}} dz, \quad b_n = \frac{1}{2\pi i} \oint_C (z - z_0)^{n-1} f(z) dz, \tag{14.45}$$

where C is any counterclockwise contour within the annular region encircling the point z_0. The result can also be combined into a single summation

$$f(z) = \sum_{n=-\infty}^{\infty} a_n(z - z_0)^n \tag{14.46}$$

with a_n now understood to be defined for both positive and negative n.

When $b_n = 0$ for all n, the Laurent expansion reduces to an ordinary Taylor series. A function with some negative power of $(z - z_0)$ in its Laurent expansion has, of necessity, a singularity at z_0. If the lowest negative power is $(z - z_0)^{-N}$ (with $b_n = 0$ for $n = N+1, N+2, \ldots$), then $f(x)$ is said to have a *pole of order N* at z_0. If $N = 1$, so that $(z - z_0)^{-1}$ is the lowest-power contribution, then z_0 is called a *simple pole*. For example, $f(z) = [z(z - 1)^2]^{-1}$ has a simple pole at $z = 0$ and a pole of order 2 at $z = 1$. If the Laurent series does not terminate, the function is said to have an *essential singularity*. For example, the exponential of a reciprocal,

$$e^{1/z} = 1 + \frac{1}{z} + \frac{1}{2z^2} + \cdots = \sum_{n=0}^{\infty} \frac{1}{n!z^n}, \tag{14.47}$$

has an essential singularity at $z = 0$. The poles in a Laurent expansion are instances of *isolated singularities*, to be distinguished from continuous arrays of singularities which can also occur.

14.8   Calculus of Residues

In a Laurent expansion for $f(z)$ within the region enclosed by C, the coefficient b_1 (or a_{-1}) of the term $(z - z_0)^{-1}$ is given by

$$b_1 = \frac{1}{2\pi i} \oint_C f(z)\, dz \equiv \mathcal{R}(z_0). \tag{14.48}$$

This is called the *residue* of $f(z)$ and plays a very significant role in complex analysis. If a function contains several singular points within the contour C, the contour can be shrunken to a series of small circles around the singularities z_n, as shown in Figure 14.6. The *residue theorem* states that the value of the contour integral is given by

$$\oint_C f(z)\, dz = 2\pi i \sum_n \mathcal{R}(z_n). \tag{14.49}$$

If a function $f(z)$, as represented by a Laurent series (14.44) or (14.46), is integrated term by term, the respective contributions are given by

$$\oint_C (z - z_0)^n\, dz = 2\pi i\, \delta_{n,-1}. \tag{14.50}$$

Only the contribution from $(z - z_0)^{-1}$ will survive—hence the designation "residue."

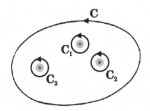

Figure 14.6 The contour for the integral $\oint_C f(z)\,dz$ can be shrunken to enclose just the singular points of $f(z)$. This is applied in derivation of the theorem of residues.

The residue of $f(z)$ at a simple pole z_0 is easy to find:

$$\mathcal{R}(z_0) = \lim_{z \to z_0} (z - z_0) f(z). \tag{14.51}$$

At a pole of order N, the residue is a bit more complicated:

$$\mathcal{R}(z_0) = \frac{1}{(N-1)!} \lim_{z \to z_0} \frac{d^{N-1}}{dz^{N-1}} \left[(z - z_0)^N f(z) \right]. \tag{14.52}$$

The calculus of residues can be applied to the evaluation of certain types of *real* integrals. Consider first a trigonometric integral of the form

$$I = \int_0^{2\pi} F(\sin\theta, \cos\theta)\,d\theta. \tag{14.53}$$

With a change of variables to $z = e^{i\theta}$, this can be transformed into a contour integral around the unit circle, as shown in Figure 14.7. Note that

$$\cos\theta = \frac{e^{i\theta} + e^{-i\theta}}{2} = \frac{1}{2}\left(z + \frac{1}{z}\right), \quad \sin\theta = \frac{e^{i\theta} - e^{-i\theta}}{2i} = -\frac{i}{2}\left(z - \frac{1}{z}\right), \tag{14.54}$$

so that $F(\sin\theta, \cos\theta)$ can be expressed as $f(z)$. Also $dz = i\,e^{i\theta}\,d\theta = iz\,d\theta$. Therefore the integral becomes

$$I = \oint_C f(z)\frac{dz}{iz} \tag{14.55}$$

and can be evaluated by finding all the residues of $f(z)/iz$ inside the unit circle:

$$I = 2\pi i \sum_n \mathcal{R}(z_n). \tag{14.56}$$

As an example, consider the integral

$$I = \int_0^{2\pi} \frac{d\theta}{1 + a\cos\theta} \quad \text{with } |a| < 1. \tag{14.57}$$

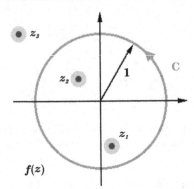

Figure 14.7 Evaluation of trigonometric integral: $\int_0^{2\pi} F(\sin\theta, \cos\theta)d\theta = \oint_C f(z)\frac{dz}{iz} = 2\pi i\,[\mathcal{R}(z_1) + \mathcal{R}(z_2)]$.

This is equal to the contour integral

$$I = \oint_C \frac{dz}{iz\left(1 + \frac{az}{2} + \frac{a}{2z}\right)} = \frac{2}{ia}\oint_C \frac{dz}{z^2 + \frac{2}{a}z + 1} = \frac{2}{ia}\oint_C \frac{dz}{(z - z_1)(z - z_2)},$$
(14.58)

where

$$z_{1,2} = -\frac{1}{a} \pm \frac{1}{a}\sqrt{1 - a^2}.$$
(14.59)

The pole at z_2 lies *outside* the unit circle when $|a| < 1$. Thus we need include only the residue of the integrand at z_1:

$$\mathcal{R}(z_1) = \frac{1}{z_1 - z_2} = \frac{a}{2\sqrt{1 - a^2}}.$$
(14.60)

Finally, therefore,

$$I = \frac{2}{ia} \times 2\pi i \mathcal{R}(z_1) = \frac{2\pi}{\sqrt{1 - a^2}}.$$
(14.61)

An infinite integral of the form

$$I = \int_{-\infty}^{\infty} f(x)dx$$
(14.62)

can also be evaluated by the calculus of residues provided that the complex function $f(z)$ is analytic in the upper half plane with a finite number of poles. It is also necessary for $f(z)$ to approach zero more rapidly than $1/z$ as $|z| \to \infty$ in the upper half plane. Consider, for example,

$$I = \int_{-\infty}^{\infty} \frac{dx}{1 + x^2}.$$
(14.63)

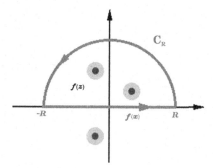

Figure 14.8 Evaluation of $\int_{-\infty}^{\infty} f(x)\,dx.$ by contour integration in the complex plane. Only singularities in the upper half plane contribute.

The contour integral over a semicircular sector shown in Figure 14.8 has the value

$$\oint_C \frac{dz}{1+z^2} = \int_{-R}^{R} \frac{dx}{1+x^2} + \int_{C_R} \frac{dz}{1+z^2}. \tag{14.64}$$

On the semicircular arc C_R, we can write $z = R\,e^{i\theta}$ so that

$$\int_{C_R} \frac{dz}{1+z^2} = \int_0^{\pi} \frac{iR\,e^{i\theta}\,d\theta}{1+R^2 e^{2i\theta}} \overset{R\to\infty}{\approx} \frac{i}{R}\int_0^{\pi} e^{-i\theta}\,d\theta = \frac{2}{R} \to 0. \tag{14.65}$$

Thus, as $R \to \infty$, the contribution from the semicircle vanishes while the limits of the x-integral extend to $\pm\infty$. The function $1/(1+z^2)$ has simple poles at $z = \pm i$. Only the pole at $z = i$ is in the upper half plane, with $\mathcal{R}(i) = 1/2i$, therefore

$$I = \oint_C \frac{dz}{1+z^2} = 2\pi i \mathcal{R}(i) = \pi. \tag{14.66}$$

Problem 14.8.1. Evaluate the following integrals:

$$\int_0^{\infty} \frac{\cos ax}{x^2+1}dx, \quad \int_0^{\infty} \frac{dx}{(x^2+1)^2}, \quad \int_{-\pi}^{\pi} \frac{d\theta}{1+\sin^2\theta}.$$

14.9 Multivalued Functions

Thus far we have considered single-valued functions, which are uniquely specified by an independent variable z. The simplest counterexample is the square root $\sqrt{z}$ which is a *two-valued function*. Even in the real domain, $\sqrt{4}$ can equal either ± 2. When the complex function $\sqrt{z}$ is expressed in polar form

$$f(z) = z^{1/2} = \left[\rho\,e^{i\theta}\right]^{1/2} = \rho^{1/2}\,e^{i\theta/2}, \tag{14.67}$$

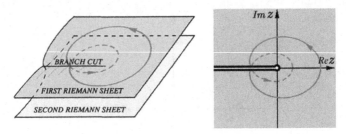

Figure 14.9 Representations of Riemann surface for $f(z) = z^{1/2}$. The dashed segments of the loops lie on the second Riemann sheet.

it is seen that the full range of $f(x)$ requires that θ vary from 0 to 4π (not just 2π). This means that the complex z-plane must be traversed *twice* in order to attain all possible values of $f(z)$. The extended domain of z can be represented as a *Riemann surface*—constructed by duplication of the complex plane, as shown in Figure 14.9. The Riemann surface corresponds to the full domain of a complex variable z. For purposes of visualization, the surface is divided into connected *Riemann sheets*, each of which is a conventional complex plane. Thus the Riemann surface for $f(z) = z^{1/2}$ consists of two Riemann sheets connected along a *branch cut*, which is conveniently chosen as the negative real axis. A Riemann sheet represents a single *branch* of a multivalued function. For example, the first Riemann sheet of the square-root function produces values $\rho^{1/2} e^{i\theta/2}$ in the range $0 \leqslant \theta \leqslant 2\pi$, while the second sheet is generated by $2\pi \leqslant \theta \leqslant 4\pi$. A point contained in every Riemann sheet, $z = 0$ in the case of the square-root function, is called a *branch point*. The trajectory of the branch cut beginning at the branch point is determined by convenience or convention. Thus the branch cut for $\sqrt{z}$ could have been chosen as any path from $z = 0$ to $z = \infty$.

The Riemann surface for the cube root $f(z) = z^{1/3}$ comprises *three* Riemann sheets, corresponding to three branches of the function. Analogously, any integer or rational power of z will have a finite number of branches. However, an irrational power such as $f(z) = z^{\alpha} = \rho^{\alpha} e^{i\alpha\theta}$ will *not* be periodic in any integer multiple of 2π and will hence require an *infinite* number of Riemann sheets. The same is true of the complex logarithmic function

$$f(z) = \ln z = \ln\left(\rho e^{i\theta}\right) = \ln \rho + i\theta \tag{14.68}$$

and of the inverse of any periodic function, including $\sin^{-1} z$, $\cos^{-1} z$, etc. In such cases, the Riemann surface can be imagined as an infinite helical (or spiral) ramp, as shown in Figure 14.10.

Branch cuts can be exploited in the evaluation of certain integrals, for example

$$\int_0^\infty \frac{x^{\alpha-1}}{1+x} dx$$

with $0 < \alpha < 1$. Consider the corresponding complex integral around the contour shown in Figure 14.11. A small and a large circle of radii R_1 and R_2, respectively, are

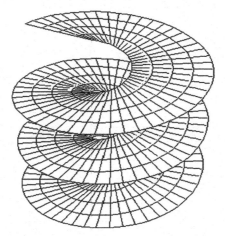

Figure 14.10 Schematic representation of several sheets of the Riemann surface needed to cover the domain of a multivalued function such as z^α, $\ln z$, or $\sin^{-1} z$.

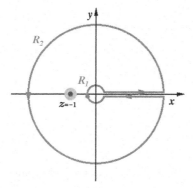

Figure 14.11 Contour used to evaluate the integral $\int_0^\infty \frac{x^{\alpha-1}}{1+x} dx$.

joined by a branch cut along the positive real axis. We can write

$$\oint_C \frac{z^{\alpha-1}}{1+z} dz = \left(\int_{R_2} - \int_{R_1} \right) \frac{z^{\alpha-1}}{1+z} dz + \int_{R_1}^{R_2} \frac{x^{\alpha-1}}{1+x} dx - \int_{R_1}^{R_2} \frac{\left[x\, e^{2\pi i} \right]^{\alpha-1}}{1+x} dx.$$

$$(14.69)$$

Along the upper edge of the branch cut we take $z = x$. Along the lower edge, however, the phase of z has increased by 2π, so that, in noninteger powers, $z = x\, e^{2\pi i}$. In the limit as $R_2 \to \infty$ and $R_1 \to 0$, the contributions from both circular contours approach zero. The only singular point within the contour C is at $z = -1$, with residue

$\mathcal{R}(-1) = (-1)^{\alpha-1} = e^{i\pi(\alpha-1)}$. Therefore

$$\left[1 - e^{2\pi i(\alpha-1)} \right] \int_0^\infty \frac{x^{\alpha-1}}{1+x} \, dx = 2\pi i \, e^{i\pi(\alpha-1)} \tag{14.70}$$

and finally

$$\int_0^\infty \frac{x^{\alpha-1}}{1+x} \, dx = \frac{2\pi i}{e^{-i\pi(\alpha-1)} - e^{i\pi(\alpha-1)}} = -\frac{\pi}{\sin(\alpha\pi - \pi)} = \frac{\pi}{\sin\alpha\pi}. \tag{14.71}$$

Problem 14.9.1.   Evaluate the integral

$$\int_0^\infty \frac{x^{-1/2}}{x^2+1} \, dx.$$

14.10   Integral Representations for Special Functions

Some very elegant representations of special functions are possible with use of contour integrals in the complex plane.

Recall Rodrigues' formula for Legendre polynomials (13.78):

$$P_\ell(x) = \frac{1}{2^\ell \ell!} \frac{d^\ell}{dx^\ell} (x^2 - 1)^\ell. \tag{14.72}$$

Applying Cauchy's integral formula (14.33) to $f(x) = (x^2 - 1)^\ell$, we obtain

$$\frac{d^\ell}{dx^\ell}(x^2 - 1)^\ell = \frac{\ell!}{2\pi i} \oint \frac{(z^2 - 1)^\ell}{(z - x)^{\ell+1}} \, dz. \tag{14.73}$$

This leads to *Schlaefli's integral representation* for Legendre polynomials:

$$P_\ell(x) = \frac{2^{-\ell}}{2\pi i} \oint \frac{(z^2 - 1)^\ell}{(z - x)^{\ell+1}} \, dz, \tag{14.74}$$

where the path of integration is some contour enclosing the point $z = x$.

A contour-integral representation for Hermite polynomials can be deduced from the generating function (13.124), rewritten as

$$e^{x^2 - (z-x)^2} = \sum_{k=0}^\infty \frac{H_k(x)}{k!} z^k. \tag{14.75}$$

Dividing by z^{n+1} and taking a contour integral around the origin:

$$\oint \frac{e^{x^2-(z-x)^2}}{z^{n+1}} \, dz = \sum_{k=0}^\infty \frac{H_k(x)}{k!} \oint \frac{z^k}{z^{n+1}} \, dz. \tag{14.76}$$

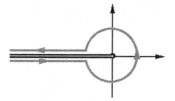

Figure 14.12 Contour for representation (14.81) of Bessel function $J_\nu(x)$ of noninteger order.

By virtue of (14.50), only the $k = n$ term in the summation survives integration, leading to the result:

$$H_n(x) = \frac{n!}{2\pi i} \oint z^{-n-1} e^{x^2 - (z-x)^2} \, dz. \tag{14.77}$$

An analogous procedure works for Laguerre polynomials. From the generating function (13.150)

$$(1 - z)^{-1} \exp\left(\frac{xz}{1 - z}\right) = \sum_{k=0}^{\infty} \frac{L_k(x)}{k!} z^k. \tag{14.78}$$

we deduce

$$L_n(x) = \frac{n!}{2\pi i} \oint (1 - z)^{-1} z^{-n-1} \exp\left(-\frac{xz}{1 - z}\right) \, dz. \tag{14.79}$$

Bessel functions of integer order can be found from the generating function (13.42):

$$\exp\left[\frac{x}{2}\left(z - \frac{1}{z}\right)\right] = \sum_{k=-\infty}^{\infty} J_k(x) z^k. \tag{14.80}$$

This suggests the integral representation:

$$J_n(x) = \frac{1}{2\pi i} \oint z^{-n-1} \exp\left[\frac{x}{2}\left(z - \frac{1}{z}\right)\right] \, dz. \tag{14.81}$$

For Bessel functions of noninteger order ν, the same integral pertains except that the contour must be deformed as shown in Figure 14.12, to take account of the multivalued factor $z^{-\nu-1}$. The contour surrounds the branch cut along the negative real axis, such that it lies entirely within a single Riemann sheet.

About the Author

S.M. Blinder is professor emeritus of chemistry and physics at the University of Michigan, Ann Arbor. Born in New York City, he completed his PhD in chemical physics from Harvard in 1958 under the direction of W.E. Moffitt and J.H. Van Vleck (Nobel Laureate in Physics, 1977). Professor Blinder has over 200 research publications in several areas of theoretical chemistry and mathematical physics. He was the first to derive the exact Coulomb (hydrogen atom) propagator in Feynman's path-integral formulation of quantum mechanics. He is the author of four earlier books: *Advanced Physical Chemistry* (Macmillan, 1969), *Foundations of Quantum Dynamics* (Academic Press, 1974), *Introduction to Quantum Mechanics in Chemistry, Materials Science and Biology* (Elsevier, 2004), and *Guide to Essential Math* (Elsevier, 2008).

Professor Blinder has been associated with the University of Michigan since 1963. He has taught a multitude of courses in chemistry, physics, mathematics, and philosophy, mostly, however, on subjects in theoretical chemistry and mathematical physics. In earlier incarnations he was a Junior Master in chess and an accomplished cellist. He is married to the classical scholar Frances Ellen Bryant and has five children. He is now a consultant with Wolfram Research, Inc. (publishers of *Mathematica*^{TM} and other technical computing products) and still pursues research in mathematical physics.

Printed in the United States
By Bookmasters